OIL, OECD, AND THE THIRD WORLD: A VICIOUS TRIANGLE?

[redacted]

OIL, OECD, AND THE THIRD WORLD: A VICIOUS TRIANGLE?

by

Hossein Askari and
John Thomas Cummings

Middle East Monograph, No. 5
Center for Middle Eastern Studies
University of Texas at Austin
Austin, Texas 78712

Library of Congress Catalogue Card Number 78-54778
ISBN 0-292-76011-6

Printed in the United States of America

CONTENTS

TABLES

Tables

ABBREVIATIONS AND ACRONYMS

ABEDA	Arab Bank for Economic Development in Africa
ADB	Asian Development Bank
ADFAED	Abu Dhabi Fund for Arab Economic Development
AFESD	Arab Fund for Economic and Social Development
CIEC	Conference on International Economic Cooperation
COMECON	Council for Mutual Economic Assistance
DAC	Development Assistance Committee
EEC	European Economic Community
FAO	Food and Agriculture Organization
GDP	Gross Domestic Product
GNP	Gross National Product
IBRD	International Bank for Reconstruction and Development (World Bank)
IDB	Islamic Development Bank
IFAD	International Fund for Agricultural Development
IMF	International Monetary Fund
KFAED	Kuwait Fund for Arab Economic Development
LDC	Less Developed Country
MSA	Most Seriously Affected (Less Developed Country)
OAPEC	Organization of Arab Petroleum Exporting Countries
ODA	Official Development Assistance
OECD	Organization for Economic Cooperation and Development
OPEC	Organization of Petroleum Exporting Countries
SAAFA	Special Arab Aid Fund for Africa
SAFD	Saudi Arabian Fund for Development

SDR	Special Drawing Right
U.A.E.	United Arab Emirates
UNDP	United Nations Development Program
Y.A.R.	Yemen Arab Republic (north)
Y.P.D.R.	Yemen People's Democratic Republic (south)

PREFACE

The oil pricing policies of the Organization of Petroleum Exporting Countries (OPEC), and especially the oil price increases of 1973-1974, have affected the world economy dramatically. In this study, we first examine the general economic impact of these oil price increases; in particular, we focus on their economic impact on the non-OPEC less developed countries (LDCs).

We conclude that the negative effect on the terms of trade has been on the order of 10 percent or less for both industrialized countries and LDCs. We analyze in detail the deterioration in the balance of trade of the LDCs during the adjustment period following the increase in the price of oil. Our arguments are specifically concerned with the consequences of a single and singular event, the 1973-1974 oil price hike, and with the reaction of the LDCs to this event. As such, future developments do not affect the significance of these findings.

Our findings indicate that substantially less than 50 percent of the negative movement in LDC trade balances between 1972 and 1975 can be traced to higher oil prices. On the positive side, both the industrialized countries (generally, the members of the Organization for Economic Cooperation and Development—OECD) and, to a somewhat greater extent, OPEC members have increased their economic support to LDCs in order to help them finance their current account deficits. This assistance, though indispensable, cannot by itself solve the long-range and fundamental economic problems of LDCs. We conclude that the industrialized countries and OPEC, rather than placing the blame for the difficulties of LDCs on each other, should *combine* their economic strengths with non-OPEC LDCs in a trilateral, cooperative approach. It is only in this way that the long-range economic future of the world's disadvantaged can be ameliorated.

In the first chapter, we discuss the background of the impact of OPEC policies and the status of the LDCs. In the second, we analyze the LDCs' balance of trade difficulties that result from increased oil prices. In the third, we examine the financial assistance given to LDCs in order to finance their increased trade deficit. In the fourth, we conclude with an evaluation of available economic policies.

Our focus is the new trilateral relationship that has emerged since 1973 among the industrially developed, oil-exporting, and non-oil LDC blocs. What will this relationship be?—trilateral cooperation or a vicious triangle?

H. A. and J. T. C.
June 1978

ACKNOWLEDGEMENTS

We are very grateful to Pat Blair, Stephen P. Magee, and to three anonymous referees for helpful and thoughtful suggestions.

Beverly Bowman has gone beyond the call of duty in typing and retyping these pages; we owe much to her patience and kindness. To Scott Lubeck, our editor, goes our thanks for his unenviable task of transforming the manuscript into a coherent book.

We also acknowledge the financial assistance of the Kozmetsky Fellowship Program and the Center for Middle Eastern Studies at The University of Texas at Austin, but as usual we are responsible for any remaining shortcomings.

1. THE SETTING OF THE PROBLEM

In the 1950s, nearly all of the members of what is now known as the Organization of Petroleum Exporting Countries (OPEC) were themselves among the poorer nations of the world.[1] OPEC was formed in 1960 in response to declining petroleum prices, both in nominal and in real terms. For the rest of that decade OPEC did little to change the futures of its members. However, beginning with the Tehran agreement of 1971 and culminating in the oil price increases of 1973-1974, many members of OPEC found themselves rapidly elevated to a wealthy though not an economically developed status.[2]

The severalfold oil price increases of 1973-1974 affected the world economy extensively. In effect, the higher prices decreed by OPEC functioned as a tax levied on the oil-importing nations. The consequence of this tax was to transfer income and wealth from oil-importing countries to OPEC; thus OPEC pushed its share of world income from about 3 percent in 1972 to about 4 percent by 1975.[3]

Increased oil prices have aggravated inflationary conditions in all oil-importing nations; ironically, the OPEC countries themselves have experienced very high rates of inflation because of the rapid infusion of money into their own economies, and because of supply bottlenecks and higher commodity prices that are partly related to higher oil prices. For the oil importers, the pressure on price levels has come both through the direct consumption of oil and through higher production costs in all sectors, though, of course, the impact has varied from country to country and from sector to sector. The "popular" view, however, is that all of today's inflation can be blamed on OPEC decisions of 1973-1974, even though the inflationary jolt of the 1973-1974 oil price increases was a onetime phenomenon and though other major factors have also been involved. Several observers have estimated the direct and indirect inflationary impact of the price hike. The International Monetary Fund (IMF), for example, calculated that it was between 2.5 and 3.5 percent for most industrialized countries, though somewhat lower in the United States (1.5 percent), and a little higher in Japan (4 percent) and Italy (5.5 percent).[4] The IMF also concluded that there were many causes for the recent rapid inflation, including the boom of 1972-1973, the large jump in many primary commodity prices, and higher oil prices. Specifically, the IMF pointed out that:

> The cost-raising influence of the oil price rise and the temporary embargo of late 1973 and early 1974 comprised *another* special feature of the recent boom that contributed to inflation.[5] (Emphasis added)

Increases in oil prices had a noticeable effect on international trade. Since the demand for oil is inelastic, oil imports did not decline sharply in response to the dramatic increase in the price of oil.[6] As a result, the trade balances of the oil importers deteriorated significantly; the deficits were even more pronounced because of the adverse impact of worldwide recession, itself partly the result of oil price increases. Coupled with the inelastic demand for oil, higher prices reduced the level of disposable income available for non-oil purchases. Thus, the effect was the same as if a tax had been imposed by the oil producers, and this mechanism explains the major part of OPEC's negative impact on the world economy. At the same time, the OPEC nations experienced a sharp positive movement in their trade balances. The deficits of the oil importers varied considerably, depending on several factors, including their dependence on imported oil; similarly, the sizes of the increased surpluses of the OPEC countries differed greatly (Saudi Arabia, rich in oil, for example, versus less generously endowed Indonesia).

In aggregate form, the broad picture can be seen in Table 1. Several points emerge from an examination of the balances of the trade and current accounts: industrial countries had a trade deficit in 1974 (a negative turnaround of $22,500 million from 1973), but by 1975 they had a surplus (a gain of $31,400 million over 1974). In

1976, they again had a deficit of $6,200 million (a decline of $27,500 million), for a drop of $18,600 million over three years. Among the more developed non-OPEC primary-producing countries, the deficit increased from $4,800 million in 1973 to $19,000 million in 1975, but then improved in 1976, declining to $16,800 million for a three year fall of $12,000 million. For the less developed countries, the deficit increased from $6,600 million to $29,100 million in 1975, but improved dramatically in 1976, declining to $15,500 million. The OPEC surplus reached a peak in 1974, declined in 1975, but increased again in 1976. Thus, although the aggregate deficit of primary-producing countries diminished sharply in 1976, especially in the less developed areas, it was still sizable. This reduction in the deficit of the less developed areas in 1976 was due to improvement in Asia and Latin America, while Africa showed little change.

In 1974, many expected that the anticipated massive shift of financial resources would prove disastrous to the world economy. Experts were predicting nearly exponential growth for the OPEC surplus. The International Bank for Reconstruction and Development (IBRD) predicted, in July of 1974, an OPEC financial accumulation (in current dollars) of $653 billion by 1980 and $1,206 billion by 1985.[7] Given such pessimistic predictions and the possibility of large "irresponsible" flows of capital, the world economy and international economic relationships were expected to meet a critical test.

However, none of these dire predictions came true, and the OPEC surplus actually declined from its 1974 high. This was largely due to an unexpectedly high rate of increase in OPEC imports. Iran's imports, for example, climbed from $3,000 million in 1973 to $18,000 million in 1975. In addition, recycling OPEC surpluses in the capital markets was achieved without excessive disturbance to the world economy. Finally, OPEC members helped in the creation of multilateral mechanisms to finance deficits of oil-importing nations.

Still lingering, however, is one major problem that the world must face. In 1974, many economists foresaw that the oil price increases would have a drastic impact on

Table 1. Global Balance of Payments Summary, 1973-1976 ($ 1,000 millions)

		Balance on					
		Trade	Services and Private Transfers	Current Account	Capital Account Balance	Change in Liabilities to Foreign Official Agencies	Balance Financed by Transactions in Reserve Assets
Industrial Countries	1973	12.4	– 1.3	11.1	–12.5	5.4	4.0
	1974	–10.1	– 1.1	–11.2	– 3.6	18.3	3.5
	1975	21.3	– 2.5	18.6	–19.2	5.2	4.6
	1976	– 6.2	4.8	– 1.4	– 3.2	15.4	10.8
Major Oil Exporters	1973	18.6	–12.4	6.2	– 2.2	—	4.0
	1974	81.7	–14.3	67.4	–21.4	0.1	46.1
	1975	52.5	–17.8	34.7	–17.3	– 0.2	17.2
	1976	63.5	–22.5	41.0	–24.0	—	17.0
Other Primary Producing Countries	1973	–11.4	1.8	– 9.6	19.5	0.1	10.0
	1974	–41.8	– 2.1	–43.8	40.6	1.8	– 1.4
	1975	–48.1	– 4.9	–53.0	45.0	5.6	– 2.4
	1976	–32.3	– 7.8	–40.1	45.7	6.4	12.0
More Developed Areas	1973	– 4.8	6.2	1.3	1.0	– 0.1	2.3
	1974	–19.1	4.8	–14.3	10.0	0.4	– 3.9
	1975	–19.0	4.2	–14.8	10.1	2.4	– 2.3
	1976	–16.8	2.5	–14.3	11.1	3.2	—
Less Developed Areas	1973	– 6.6	– 4.4	–10.9	18.5	0.1	7.7
	1974	–22.8	– 6.8	–29.5	30.6	1.4	2.5
	1975	–29.1	– 9.1	–38.2	34.9	3.2	– 0.1
	1976	–15.5	–10.3	–25.8	34.6	3.2	12.0

Table 1. Global Balance of Payments Summary, 1973–1976 (Continued)

		Balance on					
		Trade	Services and Private Transfers	Current Account	Capital Account Balance	Change in Liabilities to Foreign Official Agencies	Balance Financed by Transactions in Reserve Assets
In Africa	1973	0.5	– 2.3	– 1.9	2.1	0.1	0.3
	1974	0.7	– 3.0	– 2.3	2.3	0.2	0.2
	1975	– 2.0	– 3.1	– 5.1	4.8	0.3	—
	1976	– 1.3	– 3.9	– 5.2	4.9	0.5	0.2
In Asia	1973	– 2.4	0.1	– 2.3	4.3	—	2.0
	1974	– 9.1	0.5	– 8.6	9.8	0.9	2.1
	1975	– 9.2	0.7	– 8.5	9.1	0.5	1.0
	1976	– 2.8	0.1	– 2.7	8.4	0.3	6.0
In the Middle East	1973	– 4.3	2.1	– 2.2	3.6	0.1	1.5
	1974	– 7.0	1.4	– 5.6	5.7	0.2	0.3
	1975	– 9.1	1.0	– 8.2	6.9	1.8	0.5
	1976	– 8.2	1.6	– 6.7	6.2	1.0	0.5
In Latin America and the Caribbean	1973	– 0.3	– 4.2	– 4.5	8.4	—	3.9
	1974	– 7.4	– 5.7	–13.1	12.9	0.1	– 0.1
	1975	– 8.7	– 7.6	–16.3	14.1	0.5	– 1.7
	1976	– 3.1	– 8.1	–11.2	15.0	1.4	5.3
Total, All Countries	1973	19.6	–11.8	7.8	4.8	5.4	18.0
	1974	29.8	–17.5	12.3	15.7	20.2	48.2
	1975	25.7	–25.3	0.4	8.4	10.6	19.4
	1976	25.0	–25.5	– 0.5	18.5	21.8	39.8

Source: IMF, *Annual Report of the International Monetary Fund* (1977), p. 15.

non-oil primary-producing countries, especially on those that are least developed, and, in fact, their trade balances have badly deteriorated. Though there were some improvements in 1976 (the deficit was only about double that of 1973), the issue remains as to how much of this negative movement in the trade balances and in the larger payments deficits of the non-oil primary-producing countries is attributable to higher oil prices.

The effect of OPEC 1973-1974 actions on world trade, itself a multilateral activity, can be viewed from two perspectives: on the one hand, as an adjustment (or balance of trade disequilibrium) problem, sometimes quite severe and facing all oil-importing countries, or, on the other, as a problem of economic welfare.[8] Adjustment problems can be corrected through basic adjustment policies or repressed through quasi-adjustment.[9] Kindleberger has noted:

> Quasi-adjustment may be called for because deep-seated corrective measures are at work but slowly, and the means of financing the transitional deficit have been stretched. Or they may be applied as a second-best policy because the appropriate corrective measures are politically unacceptable, or the monetary and trade authorities lack an adequate diagnosis of the problem.[10]

For the LDCs, adjustment difficulties have a twofold nature—one direct, the other indirect. Direct adjustment problems have been felt through oil imports at higher prices, although this has been partly mitigated by the growth of LDC exports to the oil producers. Indirect adjustment problems have been caused by price increases in the exports of the industrialized countries, increases that are related to the energy inputs for these commodities. This indirect impact has also been felt as a result of the recession of 1975-1976 in the industrialized countries, which was itself brought on in part by higher oil prices, and which in turn reduced industrialized-country imports from the LDCs.

Beyond these adjustment problems, the higher price of oil (or the tax imposed by OPEC on the rest of the world) has had a negative welfare effect on the importing countries. Again, this effect has manifested itself in two ways:

directly, through higher oil prices; indirectly, through both higher prices and reduced demand (as a result of budget constraints) for non-oil goods. The terms of trade have shifted dramatically in favor of the oil exporters and to varying degrees against the importers.

Here, our principal focus will be on the *adjustment* difficulties of the LDCs since 1973. We will, however, first briefly consider the welfare problem.

The negative welfare effects of the oil price increases can be approximated by adverse movements in the terms of trade. To the extent that the terms of trade are accepted as welfare indicators, both direct and indirect aspects of the problem are included in this one measure.

However, too heavy a reliance on changes in the terms of trade as an indicator is open to question. In the first place, factors other than oil prices are involved, such as the major boom from 1972 to 1974 in the prices of many basic commodities other than oil, which clearly favored a large number of LDCs. Looking at the situation in mid-1973 from a historical perspective, one might have expected the boom to subside by late 1974 or early 1975, even *without* the OPEC actions. It is also possible that some of the most important raw materials importing countries (the OECD members in particular) expanded their money supplies partly in response, first, to the general commodity boom, or, later, to the increased oil prices, thus pushing their own export prices upward. Their motivation for such actions should be clear, at least as far as oil is concerned. The new oil prices were established in *fixed* dollar terms; higher OECD export prices would mean lower *real* prices per barrel of oil.[11]

Finally, welfare is not only a function of the terms of trade but also of the volume of trade. Thus, at best, the terms of trade are only a rough measure of the changes in welfare following the OPEC price increase.

Though this suggested welfare measure has its faults, an examination of the recent movement in the terms of trade for the industrialized countries, the oil exporters, and the non-oil LCDs may be useful (Table 2). Aggregate categories must suffice, since for most LDCs individual import indices are not yet available for 1975 or, in many cases, even for 1974.

The favorable shift in LDC terms of trade from 1972 to 1973, a jump of around 8 percent, reflects the boom in basic commodities. In 1974, despite higher oil prices (and at least some of their secondary effects on the energy-intensive exports of the industrialized countries), LDC terms of trade only dropped back to about what they had been in 1972, while those for the industrialized countries declined about 11 percent from 1973 levels. But if the latter were hit sooner and harder by declining terms of trade, their recovery was underway in 1975 with a jump of about 3 percent. On the other hand, the drop for the LDCs intensified in 1975 as the commodity boom passed and the indirect effects of higher oil prices (in addition to other factors contributing to more expensive industrialized country exports) were more fully felt; the decline from 1974 to 1975 was about 9 percent. For 1976, preliminary figures show a rebound for the LDCs to 90.4, more than 7 percent above the 1975 level. For the industrialized countries, however, there was very little change in their terms of trade over 1975.

Table 2. Terms of Trade Shifts, 1970–1976
(1970 = 100)

	1972	1973	1974	1975	1976
Industrialized Countries	100.9	97.8	86.8	89.7	89.8
Non-Oil LDCs	92.8	100.7	92.3	84.2	90.4
Oil Exporters	120.5	144.2	378.1	374.8	379.5

Note: Ratios are of export to import price indices. Categories of countries are slightly different than those used elsewhere in the text (see p. 16).

Source: *International Financial Statistics* (January and October 1977). The terms of trade were calculated from the export and import price indices. For the import price index of oil exporters for 1976, the August figure was used.

All in all, if the relative stability of 1972 is a criterion, it is apparent that the jump in oil prices, as much as it was, brought on a similarly large change in the terms of trade only for the oil producers themselves. By the end of 1976, LDC terms of trade were well within 3 percent of their 1972 levels. Even for the industrialized countries, doubly hit by increases in oil and other raw materials, the overall effect during this period, perhaps the most economically disturbing since the early years of the Great Depression, was only about 11 percent.

Moreoever, if the overall terms of trade included the price of exported services, the deterioration indicated above would be even less for the LDCs and, in particular, for the industrialized countries. In the case of an LDC such as Korea, exports of construction services to OPEC members have increased dramatically. For others, such as the Philippines, India, and Pakistan, the export of professional personnel (whose earnings are largely remitted home) to OPEC has expanded. For industrialized countries, expansion in this service sector has occurred along a broad front. Given the high inelasticity of supply of such services, the prices of these exports have in all likelihood increased more rapidly than those of exports in general.

A firmer assessment of post-1973 movements in the terms of trade of these three groups of countries must await an evaluation of the prices of traded goods during a period of economic recovery in the industrialized countries. Still, to the extent that the terms of trade can encompass the overall question of economic welfare, it is clear that, as upsetting as the 1972-1976 period may have been, the net changes in welfare for the LDCs have not been very large; in fact, from the figures shown in Table 2, the decline for the LDCs between 1970 and 1972, when oil prices varied little, was about 2.5 times the net drop from 1972 through 1976.

This admittedly crude measure of welfare changes is not, however, our central point of interest. Rather, we want to explore in some detail the impact of the 1973-1974 oil price increases on the trade accounts of LDCs—that is, their *adjustment* difficulties. In large part, this study is motivated by the often heard claim that LDC trade balances have been largely ruined by OPEC actions. This claim clearly implies that the oil drain on foreign exchange has hampered plans for economic development and has simultaneously led to huge external debts and the ensuing problems of servicing such debts. In this regard we want to consider and answer two questions. First, to what extent is this overall LDC balance of trade problem due to OPEC actions? Second, is the adjustment problem a general one or is it concentrated in a few developing countries? In addition, we will examine the financial assistance of developed countries and OPEC aimed at relieving some of this potential burden. Finally, we will discuss policy options for helping LDCs.

Before we proceed with the estimation of the contribution of oil price increases to the adjustment difficulties of LDCs, we should stress a few fundamental points. First, in using the word "contribution" we do not intend to convey the idea that there is any way to compute exactly how much or what the oil-exporting or industrialized countries should now do for the LDCs. More importantly, we should add that adjustment difficulties are also important in short- and long-run welfare considerations. The adjustment difficulties that have arisen from this implicit OPEC tax, like any other tax, have welfare implications.[12] However, the welfare impact differs from country to country, depending on a host of considerations, one of which is the internal economic policies adopted by the affected country. We are not endeavoring to consider the range of welfare effects, but rather to guage (approximately) the adjustment difficulties and to explore policies to alleviate them.

Notes

1. Current members of OPEC are Algeria, Ecuador, Gabon, Indonesia, Kuwait, Libya, Nigeria, Qatar, Saudi Arabia, the United Arab Emirates, and Venezuela.
2. The rise of OPEC has been treated elsewhere; for example, Hossein Askari and John Thomas Cummings, *Middle East Economies in the 1970s: A Comparative Approach.*
3. See p. 44.
4. International Monetary Fund (IMF), *Annual Report of the International Monetary Fund* (1975), p. 9.
5. Ibid., p. 7.

6. Any drop in oil consumption after 1973-1974 below actual or projected levels should, of course, be explained in terms of both price and income effects. Though oil demand is considered to be both price and income inelastic, the degree of price inelasticity seems to be greater.
7. As quoted in Thomas D. Willett, *The Oil-Transfer Problem and International Economic Stability*, Princeton University Essays in International Finance, p. 5.
8. Although adjustment and welfare problems are often so intertwined that it is hard to speak of one without including the other, it is worth remembering that an increase in a country's trade deficit *per se* does not mean a loss in welfare. Deficits, for example, may indicate a preference for present rather than future imports, or for domestic consumption rather than foreign sale of locally produced items.
9. See John H. Williamson, *The Crawling Peg*, Princeton University Essays in International Finance.
10. Charles Kindleberger, *International Economics*, p. 495.
11. This possibility is somewhat corroborated by the export prices of the industrialized countries. The index of export prices for these countries jumped from 100 in 1970 to 136 in 1973, and by mid-1977 it was at 207 (see *International Finance Statistics*, December 1977). In addition, there was in general a more rapid expansion of money supply from 1973 on. However, this set of policies may be attributable to domestic considerations, such as the onset of recession. This is a fruitful area for future research.
12. Clearly it is difficult, if not impossible, even to make tentative welfare judgments about the relative impact of OPEC and industrialized countries on LDC adjustment problems. For instance, if 50 percent of the difficulties of LDCs since 1972 can be attributed to OPEC and the remaining 50 percent to industrialized countries, what does this mean from a welfare standpoint? To answer this question, we would need to take into account not just the *value* of imports of LDCs from OPEC and industrialized countries, but also the *quantity* of each item. But even having done this, we have to compare welfare gains, for example, from a ton of steel versus those from a barrel of oil—at best a highly risky task.

2. BALANCE OF TRADE OF THE LESS DEVELOPED COUNTRIES

The group of non-OPEC developing nations considered here consists of 103 countries in various stages of economic development (Appendix I, Table A.1). As a general rule, though arbitrary, we excluded all net oil exporters, though several do not qualify for OPEC membership. Even though these minor exporters have been severely buffeted by the secondary effects of oil price increases,[1] examining their direct trade with the major oil exporters would serve no real purpose in indicating their difficulties. We also excluded the centrally planned economies for which price indicators did not seem to be appropriate measures for policy recommendations in 1978.[2] Several members of OECD were included in the sample (Greece, Portugal, Turkey, and an associate member, Yugoslavia), using for these OECD countries the following criteria: annual per capita income (1975) of less than $2,000 and a percentage of agricultural labor in the total labor force exceeding 25 percent.

Collectively, the countries in the sample had more than 1,500 million people in 1973, with living standards (approximated by World Bank estimates of per capita income) that ranged that year from $70 to $5,000. Some observers might quarrel with the inclusion of a half-dozen or so states at the upper end of this spectrum, such as Yugoslavia, Israel, or Singapore, but our intentions are conservative. Even the most economically fortunate of these countries have graduated so recently from LDC status that an economic upheaval as severe as the 1973 oil price realignment could threaten them with regression.

Though each country, regardless of size, has serious enough problems of its own to warrant specific consideration, it must be kept in mind that the 8 largest of the states in our samples have about 1,000 million people, approximately two-thirds of the population of the entire group. Those 25 countries with more than 10 million people have almost 90 percent of our LDC population. This emphasis is vital in the final section of this study where we discuss policy recommendations for dealing with present LDC trade-related problems.

DATA LIMITATIONS AND METHODOLOGY

Our aim is to obtain estimates, as *current* as possible, of the adjustment difficulties arising from the oil price increases for a wide cross section of developing countries; given existing data limitations, this requires some simplifying assumptions. The most current trade estimates are found in *Direction of Trade*,[3] but these are disaggregated by country, not by commodity. Thus, the imports of individual developing countries from the oil exporters can be found, but their imports of oil cannot; changes in the value of their imports can be identified, but changes in the quantity cannot. These limitations have to be accepted so that we can establish comparable estimates as recent as 1975 for a wide cross section of LDCs.[4] Our difficulty in identifying imports of oil are not too serious, at least in the aggregate. Even before the 1973 price increases, a high proportion of the value of the exports of oil producers was realized from oil or oil products; for most of these countries (Table 3), non-oil exports comprise less than 10 percent of the total value. In most cases, a closer inspection of the trade between individual LDCs and oil exporters indicates little or no involvement of non-oil commodities among the LDC imports.

Another aspect of the same problem arises from the fact that some LDCs do not obtain the bulk of their petroleum from our group of oil exporters. Despite the preponderance of OPEC members in world oil trade and despite their recent heavy investment in all phases of the petroleum and petrochemical industries, they still have relatively little refinery capacity. For example, in 1975 OPEC oil production averaged 27.1 million barrels per day, but their refinery input capacity was only about 4.4 million barrels per day.[5] Much of the latter was earmarked for local consumption; consequently most exports were in the form of crude oil. This crude is shipped to refineries in most developed and larger developing countries, from whence final petroleum products enter the consuming markets. Even a small refinery, however, is an expensive project, and many developing countries with limited development capital or small domestic markets or both do not have one yet. An examination of the trade figures for such countries indicates little direct trade with the oil-producing states, either before or after 1973. In fact, many LDCs (generally with populations of less than

Table 3. OPEC Non-Oil Exports, 1973–1975

	1973		1975	
	Total Exports ($ millions)	Percent Non-Oil	Total Exports ($ millions)	Percent Non-Oil
Algeria	1,460	24.7	3,920	8.9
Ecuador	410	75.6	835	56.2
Indonesia	2,810	57.3	5,655	33.6
Iran	5,090	11.6	20,580	3.9
Iraq	1,810	6.1	7,810	2.6
Kuwait	2,130	10.8	8,740	8.0
Libya	2,300	*	5,200	1.9
Nigeria	3,020	20.5	7,625	10.5
Qatar	410	2.4	1,760	0.6
Saudi Arabia	5,520	0.4	27,120	0.1
U.A.E.	1,240	3.2	6,810	7.3
Venezuela	3,375	11.1	8,750	5.9
Total OPEC	29,475	14.5	104,805	6.1

Notes: Total exports equal government revenues from oil exports plus non-oil exports; oil exports do not include other hydrocarbons.

*Less than 0.1 percent.

Source: *International Petroleum Encyclopedia* (1976).

5 million) get most of their oil from other non-oil producing LDCs that have refineries. This problem is confined to certain areas. In the western hemisphere, many smaller countries around the Caribbean get their oil from two non-producers with huge refinery complexes—the Bahamas and the Netherlands Antilles. These two states export little except petroleum products, and they are, for the purposes of this study, included with the group of oil producers, and arbitrarily treated as extensions of the oil-exporting countries whose oil they refine. Thus, a country that uses, for example, Venezuelan oil processed in the Netherlands Antilles can be analyzed fairly accurately here, though its trade figures show little direct trade with Venezuela or other actual oil producers.

Other regional refining centers have more varied exports. In particular, Singapore is a major exporter of a wide range of commodities, including refined petroleum, to neighboring LDCs. It is included in the list of 103 LDCs, along with two smaller refiners, the Ivory Coast and Senegal, which supply much of the refined petroleum products consumed by the francophone states of West and Central Africa. Thus, the *aggregate* oil import data for the LDCs includes most of the oil eventually consumed in the states without refineries. A few problems still remain with petroleum exports from refineries in developed states—for example, exports from Europe and Australia to present and former dependencies. However, such shipments of oil are quite small, and this discrepancy is ignored in our analysis.

Not so small are the exports of the Soviet Union, the only major oil producer and exporter among the developed countries. Soviet output grew from about 400 million tons in 1972 to 490 million tons in 1975, while exports climbed from 107 million tons to 130 million tons. During this period about 56 percent went to members of the Council for Mutual Economic Assistance (COMECON) and about 35 percent to developed countries. The rest, about 10 million tons a year, was sold to the LDCs included in our sample. Soviet sources provided about 3 percent of total LDC oil imports, which were close to 350 million tons a year during this period. Most Soviet oil, however, went to a handful of LDCs, and thus the larger role the Soviet Union plays in supplying these countries ought to be noted. Yugoslavia was the biggest purchaser, taking an average of about 4.1 million tons in 1974 and 1975, which came to about 45 percent of its oil imports. Greece was next with 1.5 million tons (16 percent of its oil imports), followed by Brazil with 1.35 million tons (5 percent). India had 1.1 million tons (6 percent), Morocco had 0.65 million tons (27 percent), and Portugal had 0.55 million tons (9 percent).

Since Soviet exports are more than 90 percent non-oil, they cannot be lumped with those of other oil exporters. Among the major LDC purchases of Soviet oil, Yugoslavia, India, and Brazil (according to *Direction of Trade*) had balance of trade surpluses *vis-à-vis* the USSR in 1975 of about $205, $170, and $70 million respectively, while Greece, Morocco, and Portugal had deficits of about $15, $28, and $100 million respectively.

In any analysis of the welfare effects of trade, a second problem also arises: the need to distinguish between *value* and *volume* of trade. The high level of worldwide inflation in recent years emphasizes the importance of this distinction. But since our basic objective is to measure the impact of the oil price increases of 1973-1974 on the balance of trade positions (that is, the adjustment-related problems) of the non-OPEC developing countries, this distinction is not directly relevant. Our major interest is how much of this change in imports and exports is due to OPEC oil price increases. If we were examining welfare implications, it would be critical to calculate what percentage of the change in exports and imports was due to *price* as compared to *volume*.[6]

The problem of calculating the indirect effects of the 1973-1974 oil price increase is another aspect of this difficulty in distinguishing between price and quantity changes. The added energy costs for producing and transporting many commodities are considerable, and to some extent this is reflected by the differences recorded in the 1974 and 1975 data.[7] In 1974 the full pressure of higher oil prices was felt only for oil and oil-derived products, but by 1975 it had hit most energy-using goods.

How much the oil price increases affected the costs of other imports varies, depending upon which country is being examined. The point here is only that, besides the "direct" impact of oil price increases, there were also "indirect" effects (that is, through third countries) of OPEC pricing policies on the trade balances of developing nations. Among these indirect effects, probably most important was that higher oil prices deepened the economic recession in the industrialized countries to such an extent that the exports of non-OPEC LDCs were below otherwise expected levels. Perhaps next in importance was the impact of oil price increases on the prices of other commodities bought and sold by the LDCs. Finally, to the extent that recession spilled over into the LDCs, their own imports declined. These indirect effects could be discussed *ad infinitum.* Although estimates of their extent are much harder to obtain than those of the direct effects, some rough approximations will be presented.

COMPUTED TRADE SHIFTS

Aggregate changes in LDC trade, relative to the shift in oil prices, are shown in Table 4. More detailed figures for the 25 largest countries in our sample are outlined in Tables A.2, A.3, A.4, and A.5. For our purposes here, the oil exporters category includes (for *all* importing countries) those so defined in *Direction of Trade*: Algeria, Indonesia, Iran, Iraq, Kuwait, Libya, Nigeria, Oman, Qatar, Saudi Arabia, United Arab Emirates, and Venezuela—all members of OPEC except for Oman. To this we added (for *all* importers) the two remaining OPEC members: Ecuador and Gabon. Selectively, for each importing country on the basis of historical trade patterns, we also added these minor oil exporters: Angola, Bahrain, Bolivia, Brunei, Colombia, Congo Republic, Egypt, Syria, Trinidad-Tobago, and Tunisia (as well as the two oil-refinery centers, Bahamas and the Netherlands Antilles).

The adverse impact of oil price increases has clearly been only part of the LDC trade-balance-adjustment problem, no matter which of the measures shown in the tables is considered. For the smaller countries not listed in the tables found in the appendices, the conclusions are similar to those for the group as a whole and for the larger states alone. (Though given the complexities of the world energy situation, individual exceptions were found using our somewhat crude criteria.) For the 25 largest LDCs, shifts in aggregate trade balances with the oil producers account for about 25.4 percent of their overall balance of trade decline between 1972 and 1975; the oil producers were the source of 18.9 percent of the group's increased import costs.

Table 4. Aggregate LDC Shifts in Trade, 1972–1974 and 1972–1975

	1972–1974	1972–1975
Absolute Change in LDC Trade Balances with:		
(1) All Countries ($ millions)	–28,375.1	–35,666.8
(2) Oil Exporters ($ millions)	–10,330.3	– 9,009.5
(2) As Percentage of (1)	36.4	25.3
Change in Imports:		
From World (percent)	125.7	133.3
From Oil Exporters (percent)	356.3	370.6
Percent of Total Increase Directly Related to Oil Exporters	18.9	18.5
Change in Exports:		
To World (percent)	98.6	91.8
To Oil Exporters (percent)	212.4	332.8
Percent of Total Increase Directly Related to Oil Exporters	7.6	12.5

Source: *Direction of Trade* (Annual Editions, 1975 and 1976).

Of course, it must be emphasized that, taken alone, neither the overall shift in trade balances nor the change in imports indicates the full extent of the adjustment problems of these countries. Indicators perhaps more frequently cited than these are balance of payments figures that include the balance of trade. But these can be misleading, especially in a period of high worldwide inflation, such as has prevailed since the late 1960s.[8]

The trade balances of the LDCs have, of course, been adversely affected by oil prices, but the export prices of the industrial countries were climbing rapidly well before October 1973. For example, according to the International Monetary Fund, these were up by 49 percent from 1968 to the end of 1973, and climbed another 36 percent above 1973 levels by the end of 1975; the average annual rate of increase during this period was more than 10 percent. Export prices for the LDCs themselves were also sharply up—by some 103 percent from 1968 to 1975 (also at an annual rate of about 10 percent) for the group as a whole (excluding the oil producers)—but some countries benefited far more than others. For example, from 1968 to 1975, export prices rose by about 205 percent for Morocco, 113 percent for Pakistan, and 128 percent for Brazil, but by only about 43 percent for Sri Lanka, less than 30 percent for Uruguay, while they fell 5 percent for Zambia.

From 1972 to 1974 the imports of all 103 LDCs climbed by about $72,100 million, about 125 percent over the 1972 level of $57,400 million. About $13,600 million of this increase can be traced directly to imports from the oil-exporting states, and it represents a jump of nearly 360 percent over 1972 imports of $3,800 million. The estimated shift in imports through 1975 is slightly larger—an increase over 1972 of $76,500 million for all imports and of $14,100 million for those originating in the oil-producing countries. The added oil bill for the 103 LDCs is certainly large in both the periods from 1972 to 1974, which was before the full effect of increased oil prices had been felt on all their imports, and from 1972 to 1975, when most of the impact was felt. However, oil is not the only major factor in these import bills. Higher payments to the oil exporters amounted to about 18.9 percent of increased import costs for 1972 to 1974 and to about 18.4 percent for 1972 to 1975. The oil exporters' aggregate share of LDC imports doubled between 1972 and 1975 to more than 13 percent. As the figures in Table A.5 show, there is quite a bit of variance in these measures within the group of larger countries, which is also the case for the smaller countries not listed. For example, the oil countries accounted for about a fifth or more of the imports of Brazil, the East African Community, Pakistan, the Philippines, and Thailand in 1975, but for well less than 10 percent of the imports of Mexico, Burma, Morocco, and Zaire.[9]

The role of the oil exporters in the cumulative balances of trade of these LDCs is larger than that indicated when we consider imports alone. From 1972 to 1974, the aggregate trade balances of these countries declined by some $28,400 million; the drop from 1972 to 1975 was nearly $35,700 million. Their trade balances with the oil producers fell by over $10,300 million from 1972 to 1974, and this was 36.4 percent of the total drop experienced by the LDCs as a group. In 1975 there was a recovery of sorts. The estimated 1972 to 1975 decline for their trade balances with the oil producers was about $9,000 million, or 25.2 percent of their overall trade balance decrease.

The relative improvement in 1975 was in part the result of a general phenomenon observed in the trade patterns of the oil states—massive increases in their imports as higher oil prices began to swell their financial reserves by mid-1974. Most of these purchases were part of new or revised development plans—for example, Saudi Arabia's 1975-1980 plan that proposed expenditures close to $145,000 million, or the Libyan proposal for 1976 to 1980 that included projects totalling nearly $25,000 million. Of course, these purchases encountered many bottlenecks; ships waited for weeks to be unloaded in inadequate ports, truck queues measured in kilometers were backed up at customs stations, and so on. In fact, many observers in late 1973 and early 1974 were concerned by the likelihood that several OPEC states would be unable to absorb huge import increases. While their concerns have been abated by the nearly insatiable appetites of these states' planners and consumers, there remains the

Table 5. LDC Trade with Developed Countries and Oil Exporters, 1970–1975

	Total		Developed Countries			Oil Exporters		
	Value ($ million)	Index	Value ($ million)	Index	Percent of Total	Value ($ million)	Index	Percent of Total
Imports								
1970	46,072.7	100.0	33,894.0	100.0	73.6	2,635.7	100.0	5.7
1972	57,638.9	124.5	41,696.3	123.0	72.7	3,807.9	144.5	6.6
1973	81,142.3	176.1	57,451.5	169.5	70.8	5,834.5	221.4	7.2
1974	129,461.4	281.0	86,205.3	254.3	66.5	17,374.7	659.2	13.4
1975	133,861.2	290.5	90,435.0	266.8	67.6	17,921.0	679.9	13.4
Exports								
1970	35,215.4	100.0	24,969.4	100.0	70.9	1,411.2	100.0	4.0
1972	44,496.0	126.4	31,493.5	126.1	70.8	1,533.5	108.7	3.4
1973	65,454.0	185.9	44,322.7	177.5	67.7	2,580.5	182.9	3.9
1974	88,377.5	251.0	59,389.2	237.8	67.2	4,790.5	339.5	5.4
1975	85,321.5	242.3	55,775.0	223.4	65.4	6,637.1	470.3	7.8

Note: LDC trade figures are the summed totals for the countries listed in Table A.1. Developed countries include North America, Japan, Australia, New Zealand, South Africa, and all western European countries not included among LDCs.

Source: *Direction of Trade* (Annual Edition, 1976).

basic question of the wisdom and long-range prospects of many of the current projects that give rise to massive increases in OPEC imports.[10]

In the aggregate, LDC exports to the oil producers swelled from some $1,500 million in 1972 to $4,800 million in 1974 and to more than $6,600 million in 1975 (a jump of about 333 percent over three years), while their total exports were up by about 90 percent (from about $44,500 to about $85,300 million). The oil producers' share of LDC exports more than doubled, from 3.4 percent in 1972 to 7.8 percent by 1975. Still, as exporters mostly of raw materials, these LDCs continued to find their major markets in the industrialized economies of western Europe, North America, and Japan, and their added imports from these developed states more than outpaced their new exports (Table 5). From 1972 to 1975, LDC imports from the developed states were up by almost $49,000 million, or about twice their $24,300 million increase in exports to these same countries, and the LDC trade deficit with the developed states grew from $10,200 million to nearly $35,000 million.[11]

For individual countries, however, it is clear that some LDCs, like some developed countries, have been notably more successful in increasing their exports to oil producers than have others. A few LDCs (Table 6) have seen these exports rise more rapidly than imports. Several of these LDCs, not surprisingly, are at the upper end of the development spectrum—Argentina, Greece, and Singapore—exporting a wide variety of commodities (as does less prosperous India) now in heavy demand in the affluent oil states. Others are fortunately situated near major oil regions—Somalia, Lebanon, Cyprus, and Malta. A few others (Table 7) have also experienced rapid increases in their exports to the oil countries, though their deficits may have widened considerably as their oil imports grew even faster. Pakistan, Brazil, and Yugoslavia, plus the countries listed in Table 6, accounted for about 80 percent of the increased LDC exports to the oil countries, but for only about half the additional LDC imports between 1972 and 1975.

Another group of LDCs has been at least partly insulated from the direct effects of the oil price hike because it produces a significant fraction of its own oil (Table A.6). Some of these states have reasonably good hopes of soon becoming self-sufficient in oil, and even of becoming modest net exporters. Thus, Mexico and Zaire have already joined, as Burma may soon, these ranks, which include Tunisia, Bolivia, and the Congo Republic. Other LDC oil producers, notably Argentina, Brazil, Peru, Chile, Malaysia, India, and Pakistan, are officially optimistic about their prospects, but they generally have less hard geological evidence to back up their projections. Still others with little or no present production are also hopeful—for example, the Philippines and several coastal states in Africa. On the other hand, at least one longtime

Table 6. Changes in Trade with the Oil Exporters, Selected LDCs, 1972–1975 ($1,000's)

	Imports		Exports	
Country	1972	1975	1972	1975
India	$ 252.1	$1,292.4	$ 137.1	$ 765.9
Mexico	61.3	124.2	45.6	111.2
Argentina	43.8	231.4	19.9	176.4
Taiwan	224.5	849.8	158.0	464.0
Greece	129.6	882.8	40.3	356.2
Hong Kong[a]	74.1	145.7	182.4	490.1
Somalia	2.0	16.0	26.5	63.3
Lebanon	62.6	112.6	158.6	746.5
Jordan	16.4	82.8	15.4	44.2
Singapore[a]	382.7	1,722.2	44.5	386.7
Cyprus	13.3	23.3	6.1	27.6
Malta	0.4	4.6	3.7	21.1
All LDCs	$3,775.1	$17,921.0	$1,530.6	$6,637.1

Note: [a] Hong Kong's imports from the oil states show little change since it obtains much of its oil from a regional refinery center, Singapore, which reexported petroleum products in 1975 worth more than 85 percent of its imported crude petroleum.

Source: *Direction of Trade* (Annual Edition, 1976).

Table 7. Percentage of Extra LDC Exports Directly Traceable to Oil Exporters, 1972–1975

More than 40%	30% to 40%	20% to 30%	10% to 20%
Bangladesh	India	Ethiopia	Brazil
Pakistan	Sri Lanka	Sudan	Turkey
Burma	Benin	Afghanistan	Argentina
Somalia		Greece	Yugoslavia
Lebanon		Mozambique	Taiwan
Cyprus		Jordan	Hong Kong
			Chad
			Israel
			Singapore
			Jamaica
			Malta
			Djibouti

Source: *Direction of Trade* (Annual Edition, 1976).

net exporter (and thus not included among the LDCs in this study), Colombia, faces dwindling output and permanent reversion to the status of a net oil importer, which it became in 1976, if sizable, new, and commercially exploitable deposits are not found soon.

But most of the LDCs have neither much promise based on past and recent export performance to OPEC members, nor much prospect of increased domestic oil production. Still, considerable distinctions among the LDCs can be made on the basis of how much of their increased import bill and/or trade balance deterioration can be directly traced to oil price increases (Tables 8 and 9). All things considered, the countries listed in the upper categories of both tables are those with the most pressing post-1973 problems. Excepting those countries with large refinery complexes that supply neighboring countries (like Singapore or Ivory Coast), the group with the greatest difficulties includes Brazil, the Philippines, Thailand, the East African Community (Kenya, Tanzania and Uganda), South Korea, Chile, Uruguay, Jamaica, Guyana, Surinam, and Barbados. Several of these countries have seen the portion of their imports not covered by exports rise more sharply than have the LDCs as a whole (Table A.7). Another group of states that places slightly lower in

Table 8. LDC Extra Import Bill Directly Traceable to Oil Exporters, 1972–1975

More than 40%	30.1 to 40%	20.1 to 30%	10.1 to 20%	0 to 10%[a]
Panama[b]	East Africa	India	Bangladesh	Mexico
Guyana	Uruguay	Brazil	Argentina	Zaire
Cyprus	Jamaica	Pakistan	Yugoslavia	Afghanistan
Swaziland		Philippines	Sudan	Morocco
Barbados		Thailand	Cent. Amer.	Peru
		Turkey	Taiwan	Y.A.R.
		South Korea	Malaysia	Dominican Rep.
		Ethiopia	Chile	Hong Kong
		Sri Lanka	Ghana	Chad
		Greece	Portugal	Lebanon
		Zambia	Mozambique	Papua New Guinea
		Singapore[b]	Haiti	Mauritius
		Liberia	Ivory Coast[b]	Reunion
		Surinam	Senegal[b]	Guadeloupe
		Cape Verde	Israel	Djibouti
		Belize	Somalia	New Caledonia
		Sao Tome	Sierra Leone	Bermuda
			Jordan	
			Y.P.D.R.	
			Martinique	
			New Hebrides	
			Fr. Guiana	

Notes:

[a] Omitted from this category are some 31 countries, nearly all with less than two million population and concentrated in Africa and the Pacific regions for which *Direction of Trade* indicates little or no trade with the oil exporters; see p. 32, no. 13.

[b] Regional supplier of refined petroleum products.

Source: *Direction of Trade* (Annual Edition, 1975).

Table 9. Percentage of Overall Negative Trade Balance Shifts for the LDCs Due to Shifts with Oil Exporters, 1972–1975[a]

More than 50%	40.1 to 50%	30.1 to 40%	20.1 to 30%	10.1 to 20%	0 to 10%[b]
Thailand	Brazil	Taiwan	India	Pakistan	Bangladesh
South Korea	Philippines	Greece	Turkey	Ethiopia	Argentina
East Africa	Cent. Amer.	Zambia	Portugal	Yugoslavia	Sri Lanka
Malaysia[c]	Bermuda	Sao Tome	Haiti	Sudan	Mozambique
Ghana			Senegal[d]	Morocco	Y.A.R.
Ivory Coast[d]			Cape Verde	Peru	Chad
Dominican Rep.[c]			New Caledonia	Israel	Jordan
Uruguay				Sierra Leone	Guadeloupe
Singapore[d]				Martinique	Djibouti
Jamaica					Fr. Guiana
Panama[d]					
Y.P.D.R.[c]					
Liberia					
Mauritius					
Surinam					
Barbados					

Notes:

[a] Not listed are those countries showing a positive shift with the world and the oil exporters, or at least with the oil exporters alone; falling into this category are: Mexico, Burma, Afghanistan, Cameroon, Hong Kong, Somalia, Lebanon, and Cyprus.

Also excluded are those showing a positive shift with the world and a negative shift with the oil exporters for both 1972–1974 and 1972 - 1975; falling into this category are: Chile, Papua New Guinea, Guyana, Swaziland, Belize, and New Hebrides.

[b] Excluded from this category are those countries showing little or no trade with the oil exporters. See note a., Table 8.

[c] 1972–1974.

[d] Regional supplier of refined petroleum products.

Tables 8, 9, and A.7 is made up of Turkey, the Central American Common Market (Costa Rica, El Salvador, Guatemala, Honduras, and Nicaragua), Taiwan, Ghana, Greece, and Zambia. In singling out these 23 countries (about one-third of those LDCs for which *Direction of Trade* figures allow us to use imports from the oil exporters as a reasonable proxy for oil imports), we have a group with some 360 million people, about 24 percent of the population of our entire sample of LDCs.

It is possible to be more optimistic about the trade balances of some of these countries than the figures suggest. For example, South Korea was, in absolute terms, hit harder by higher post-1973 import bills than any other LDC except Brazil, which has more than three times Korea's population. The Korean trade deficit rose from about $900 million in 1972 to about $2,300 million in both 1974 and 1975, as imports jumped from $2,500 to $7,300 million between 1972 and 1975. But by the end of 1976, indications were that the deficit was cut to about $1,000 million, as exports climbed more than 50 percent over the 1975 level. A major part of the Korean accomplishment can be traced to a successful export drive among the oil exporters at a time (1976) when several OECD members were imposing tighter quotas on major Korean exports. According to IMF estimates, Taiwan restored its trade surplus in 1976 after two years of deficits, as exports gained more than $2,800 million over 1975 levels. In Central America, Guatemala began commercial oil production in 1976 (Table A.6), which may soon make the largest member (with about a third of the population) of the Central American Common Market self-sufficient, or even a net exporter.[12]

A few countries not included among the LDCs are in still another category because they are themselves oil producers, but have net oil exports that are small in relation to their overall economies. Like other LDCs, however, they are subject to the secondary effects of oil prices, which may be considerable. In particular, Egypt, Syria, Tunisia, Bolivia, and Colombia, with an aggregate population of more than 75 million in 1973, should be mentioned. The trade profiles of these countries for the 1972 to 1975 period are indicated in Tables A.8 and A.9.

The *Direction of Trade* data used here disaggregate trade only by *country*, not by *commodity*. This gives rise to a last group of LDCs. Not much can be said in this context about this group because its members carry on little trade *directly* with the oil-producing states. These countries, mostly in central Africa and in the Pacific region, get their oil from third parties—regional refinery complexes that, in most cases, are located in the LDCs (like those in Singapore or the Ivory Coast). They number 31, slightly less than a third of our LDC sample; however, they have only about 70 million people, about 4.5 percent of the total LDC population.[13]

Including this group in the study, however, helps us to account for the oil bought by several LDCs for processing and reexporting. Moreover, the group's aggregate world trade contributes to an understanding of the overall trade problems of the LDCs in the 1970s. Several of these countries, particularly those in Africa, have trade deficit-to-export ratios quite a bit above the LDC average (Table A.7), and had notable jumps in this ratio between 1972 and 1975. It can be assumed that many of these countries are among those with the greatest adjustment problems as a result of recent world trade events.

Up to this point, discussion has centered around the direct or primary effects of the oil price increase. However, oil price increases deepened worldwide recession, which in turn reduced the exports and imports of LDCs from what they would have been; thus, there are indirect or secondary adjustment effects or problems. For example, the rise in oil prices may have hiked the prices (but not the profitability) of LDC exports since oil is an input in many products, or, by adding to transport costs, this rise could have increased CIF prices, thus (depending upon elasticities) possibly reducing net receipts from these exports. Also, depending upon elasticities, it is possible that (from the restricted viewpoint of LDC trade balances) the oil price increase could have produced "beneficial" secondary adjustment effects: the higher export prices of developed nations might have reduced the expenditures of LDCs on imports from what they might have been in the absence of such higher prices.[14] The extent of these effects depends upon the size of the appropriate income and price elasticities, and upon the scope of both the deepening recession in industrial countries and the worldwide inflation that have resulted from OPEC actions.

Calculating the indirect effects depends, first of all, upon gauging the impact of oil price increases on world income and on the export prices of both the non-OPEC LDCs and the industrialized countries. Then, if income and price elasticities of demand for imports can be estimated, the magnitude of these indirect effects can be discussed.

One recently available estimate for the weighted income elasticity of demand in industrial countries for LDC exports has been derived from rather detailed analysis; the author, Magee, presents a fairly high value—1.3.[15] Following this course, the next step—identifying the magnitude of the decline in income in the industrial countries resulting from OPEC oil price increases—is difficult.

However, an upper bound for this drop can at least be approximated. During the period 1960 to 1970, the average annual change in real GNP in the industrial countries was 4.8 percent. But in 1974 the corresponding figure was 0 percent; in 1975 it was -1.5 percent. At the worst, a decline of 4.8 percent in 1974 to 11.3 percent in 1975 (that is, 4.8 percent compounded for two years, plus 1.5 percent, which was 1975's negative growth) in real GNP (relative to what could be projected historically from 1973 levels) can be attributed to OPEC activities. These figures are *upper-bound* limits—it must be emphasized that the recession in the industrialized countries, which began in 1974, had other roots in addition to higher oil prices. A more scientific estimate is provided by Richardson:

> The difference between a $7 and $11 world oil price might be in the region of 0.5 percent in the GNP growth rate.[16]

Given the impact in the price of oil in 1973-1974, the adverse impact on GNP growth rates would be on the order of 1 percent. Multiplying Magee's 1.3 income elasticity estimate by 1 percent (for 1974) and 2.02 percent (1 percent compounded for 2 years, for 1975) gives values of 1.3 percent for 1974 and 2.63 percent for 1975. In other words, LDC exports to the industrialized countries would have been some 1.3 percent (1974) to 2.63 percent (1975) higher had it not been for the recession in the industrialized countries. Thus, our upper bound is a reduction in exports—1.3 percent of the 1974 figure in 1974, 2.63 percent of the 1975 figure in 1975—and amounts to $767 million in 1974 (1.3 percent of $59,389 million) and $1,489 million in 1975 (2.63 percent of $56,631). These figures should now be expressed in terms of trade balance deterioration so that they can in turn be added to the direct impact. Thus, in 1974, the indirect impact on the trade balance was about 3 percent ($767 million divided by $28,375 million); for 1975, it was 4 percent ($1,489 million divided by $35,667 million). Again we must emphasize that this is an order of magnitude estimate of the indirect effect.

The negative impact of oil price increases on LDC exports, due to rising prices in LDCs, could be similarly estimated—using the price elasticity of demand for LDC goods in the industrialized countries in conjunction with the increase in LDC export prices. However, a recent study made in this regard by Houthakker and Magee does not indicate statistically significant *negative* and non-zero price elasticities.[17] In fact, some are positive. Thus, we have omitted this effect from our overall estimates.

Since our concern is to estimate the *major* individual effects, we have also chosen to neglect the impact of the oil price increases on recession in LDCs and thus on reduced LDC imports. Altogether, it would appear that in 1974 a total figure of about 29 percent of the LDC trade balance deterioration could be directly (26 percent) and indirectly (3 percent) attributed to OPEC. For 1975 this figure was 27 percent (23 percent direct and 4 percent indirect). We must stress the roughness of our estimate of the indirect as compared to the direct component of these estimates. But it must be remembered that our indirect measure is an order of magnitude estimate. Thus we can state emphatically, even if we allow for other large effects that were not included in our estimates,[18] that *substantially* less than half of the deterioration of LDC trade balances can be traced to OPEC price increases in 1973-1974.

Up to this point, we have concentrated on the negative adjustment effect of OPEC oil price increases on LDCs. However, OPEC members, along with the industrialized countries, have granted financial assistance aimed at easing this adjustment difficulty for the LDCs. In the next chapter, we will examine such efforts.

Notes

1. Some indications of this for five smaller exporters can be seen on p. 27.
2. Centrally planned developing economies excluded are those in Asia, (China, Mongolia, North Korea, Vietnam, Laos, and Cambodia), Latin America (Cuba), and southeastern Europe (Albania, Bulgaria, and Romania). However, Yugoslavia was included in the sample because of its strong ties to the market economies.
3. International Monetary Fund, *Direction of Trade* (Annual Issues, 1969-1975 and 1970-1976).
4. Other sources of trade data disaggregate trade data by commodity and trading partner. *Commodity Trade Statistics,* for example, uses such a format, but only for about 40 LDCs, and often with a considerable publication lag. *The Yearbook of International Trade Statistics* presents figures for as wide a cross section as we used here, but the latest currently available estimates are pre-1974 for many LDCs. The IMF reports estimates for crude oil imports in *International Financial Statistics,* but only for selected LDCs.
5. *International Petroleum Encyclopedia* (1976).
6. In fact, these LDCs imported about the same amount of oil in 1974 and 1975 as in 1973; see below, fn. 9, for this chapter.
7. *Direction of Trade* data quoted here are somewhat more accurate for 1974; 1975 figures for several countries are estimated, based on extrapolations from partial annual data.
8. Balance of payments data, as collected and presented by the International Monetary Fund in the *Balance of Payments Yearbook,* are not available for more than a third of the countries in our sample. In addition, the latest available data for many LDCs are for 1973.
9. Of course, it must be emphasized again that these figures represent only monetary and not quantitative changes. Though price increases after 1972 were sizable for many major import items, particularly for capital goods and cereals, there were quantitative increases in the LDC imports of many goods. However, the 1973 oil price hike put constraints on the oil imports of the LDCs. For example, one authoritative source, the 1976 edition of the British Petroleum Company's *Statistical Review of the World Oil Industry,* estimated that oil imports by developing countries showed the following recent trend (in millions of tons): 1970—245; 1971—297; 1972—326; 1973—347; 1974—339; 1975—337. But in 1976, LDC imports rebounded sharply to 396 million tons, a 17.5 percent jump over 1975 import levels and 14.1 percent above the previous peak in 1973. By way of comparison, 1976 oil imports in the industrialized countries were also up notably over 1975, registering an 11.8 percent average gain to 1,309 million tons, but this level was still 2.7 percent below their peak import year, 1973.
10. The authors have discussed this question in another study: *Middle East Economies in the 1970s: A Comparative Approach.*
11. The dominant role the industrialized countries play in LDC trade is further emphasized when we look at how much the LDCs trade with each other. In

1975, LDC exports to other non-oil exporting LDCs amounted to about $16.6 billion (about 19.1 percent of their total exports), while their imports from each other came to $18.5 billion (13.6 percent of total imports). Intra-LDC trade has been growing slightly faster than trade between LDCs and the industrialized countries. The former climbed by 161 percent on the export side and about 179 percent on the import side between 1970 and 1975.

12. Clearly trade must be seen as a multilateral activity and, as such, increased exports from Korea, for example, to third markets can compensate for the bilateral deterioration *vis-à-vis* OPEC. However, we are primarily interested in the negative contribution of OPEC oil price increases on the overall change in the balance of trade.
13. The group consists of Nepal, Madagascar, Upper Volta, Mali, Guinea, Malawi, Niger, Rwanda, Burundi, Benin, Paraguay, Togo, Central African Empire, Mauritania, Lesotho, Namibia, Botswana, Fiji, Guinea-Bissau, Gambia, Malta, Equatorial Guinea, Comoros, Macao, Solomon Islands, Western Samoa, French Polynesia, Tonga, Gilbert and Ellice Islands, Seychelles, and Nauru.
14. Of course, in this context, such a benefit refers only to adjustment and not to welfare issues.
15. This estimate is taken from Stephen P. Magee, "Tariff Preferences for Less Developed Countries, World Wage Income and Welfare," (Background paper prepared for the United Nations—International Labor Organization World Conference on Employment, Income Distribution, and Social Progress in the International Division of Labor).
16. Harry W. Richardson, *Economic Aspects of the Energy Crisis*, p. 7.
17. H. S. Houthakker and S. P. Magee, "Income and Price Elasticities in World Trade," *Review of Economics and Statistics*, Vol. LI (May 1969), p. 113.
18. There is, for example, the problem of estimating how much the prices of LDC non-oil imports rose as a result of oil price increases and how the higher prices of the non-oil imports affected the quantities LDCs purchased in 1974 and 1975. We indicated above (page 2) that the IMF estimated the onetime effect of the 1973 oil price hike on the price levels of industrialized countries to have been between 2.5 and 3.5 percent for most countries. If we used this as an approximation for the effect on industrialized country exports to LDCs as well, we would then need to know both price and income elasticities in the LDCs for these goods in order to estimate the effect of oil prices on LDC non-oil import bills.

3. INTERNATIONAL FINANCIAL ASSISTANCE TO THE LESS DEVELOPED COUNTRIES

Our principal focus must be on the bigger countries. The 25 largest LDCs, those with more than 10 million people (Table A.1), have an aggregate population that is nearly 90 percent of that of the group as a whole. It is primarily the difficulties of these larger states that must be faced if any serious attack is to be made on LDC trade problems. Of the aggregate $48,500 million deficit in the 1975 trade of these 103 LDCs, nearly 10 percent can be traced to Brazil *alone*; about 30 percent more was accounted for by the deficits of Mexico, Yugoslavia, Turkey, the Philippines, and Thailand. Mexico (whose 1975 imbalance on a per capita basis was nearly 50 percent worse than Brazil's) accumulated this deficit even though it was nearly self-sufficient in oil and thus did not feel the first-order effects of increased oil prices. In 1976 and 1977, Mexico graduated to the ranks of the smaller oil exporters and would thus by now be removed from our group of oil-importing LDCs.

The examples of Brazil and Mexico illustrate the need for examining the current trade-related problems of the LDCs on a more particular, less inclusive level than the aggregate figures seen above, whenever future policy considerations are involved. Mexico may have been subject to some buffeting since 1974 by the secondary effects of higher oil prices, but with a 1975 trade deficit equal to about 125 percent of its total exports, meaningful solutions to its problems must be based on probing more than its foreign trade sector alone.[1]

Mexico has long been approximately self-sufficient in oil. Brazil on the other hand produces only about a quarter of what it consumes (Table A.6). While both Brazilian and foreign observers have frequently commented during the past decade on the country's vast potential for oil and gas production, actual gains in both production and additions to confirmed reserves have hardly been outstanding. In fact, this earlier optimism may have been unfounded, but another explanation for stagnation in the domestic petroleum industry may be found in Brazil's general reluctance to allow exploration by foreign companies.[2] Without denying that this Brazilian aversion may be based on reasonable political decisions, it may well be that the recent trade problems of Brazil (and perhaps other LDCs), which have been blamed on higher oil prices, may be largely explained in terms of political decisions made (or not made) by these countries *themselves* before and since the OPEC price increases.[3]

Although we have given most of our attention to the larger countries, we do not intend to ignore the very severe difficulties faced by many of the smaller LDCs—most of which have a much more limited range of exports than countries like India, Brazil, Mexico, Turkey, Korea, or Taiwan. However, the remedies prescribed during the last few years for LDC trade disequilibria, such as assistance in the form of grants or concessionary loans, can hardly be followed by many of the larger LDCs. For example, Brazil and Mexico are already the largest debtors among the LDCs; Zaire was perilously close to default on its debt several times in 1975 and 1976, and, as mentioned above, Peru was being watched closely by bankers at the end of 1977.

On the other hand, if the potential grantors and creditors are willing, the traditional remedies can be used a little longer for many of the smaller countries. Their problems, while often enormous relative to their own economies, can be measured in tens, not thousands, of millions of dollars. For example, the situations of 15 to 20 of the poorest LDCs, stretching in a wide band across Central Africa, are typified by Mali, Niger, and Rwanda, whose 1975 deficits were about $180, $90, and $53 million respectively. Subsidizing the smallest and poorest nations would cost the rich (including several of the oil exporters) $5,000 to $8,000 million a year—a sum (in current prices) they could afford indefinitely. But even the largesse of the most generous donors and the most understanding lenders would soon be exhausted if the widening trade deficits of Brazil, Mexico, Turkey, the

Philippines, Thailand, East Africa, Yugoslavia, or Greece have to be financed at these levels into the 1980s.

Since 1970, in the face of both rampant inflation in import prices and the need for quantitative increases in many of these imports (occasioned by events such as the Sahelian drought of the early 1970s), economic assistance of various types has been a major factor in the economic viability of many LDCs. Though, as can be seen from the trade figures of the last six years or so, the LDC trade deficits did not originate in, nor were they widened since 1973 solely (or even mostly) by the oil price hikes of 1973-1974, this singular event serves as a watershed in discussing recent external assistance given to the LDCs. Much of this aid was prompted by the sudden jump in import bills that followed the price increases, and, in fact, a good part of it has been granted by the oil exporters themselves.

Both the financing of the negative trade impact of increased oil prices and the role of various countries and institutions in this process are important to an appreciation of the total picture. To the extent that OPEC members financed the LDC deficit in the form of grants of concessionary loans, the adjustment difficulties due to oil price increases have been reduced.

To finance their current deficits of more than $50,000 million in 1975, the non-oil LDCs received funds from many sources. They borrowed nearly $3,000 million from the IMF and $12,500 million from private banks; they also obtained $14,500 million (net) from official development assistance (ODA). Another $15,500 million was received in trade credits, official non-concessional loans, and private direct investment. About $3,500 million of their own financial reserves were used.[4]

Much of this financing of current annual deficits, however, has been concentrated among the "middle-income" group of LDCs, that is, those with per capita incomes ranging from $200 to $700. The lowest income group (less than $200) has faced very severe foreign exchange constraints and thus low rates of growth—even, in some cases, negative growth rates. The OECD, in comparing the economic assistance of OPEC members with those of the western countries that are members of the Development Assistance Committee (DAC), said:

> Private capital (direct investment, Eurocurrency borrowing and private export and other credit) amounted to almost $32 billion, mainly channelled to the higher-income countries, which also received the bulk of official export credits and non-concessional loans from international financial institutions, a further $9 billion. The higher-income countries may also have received more than their proportional share of the $17 billion of official development assistance extended by DAC and non-DAC donors. Of this amount, $4 billion was channelled to multilateral agencies to be redirected, mainly to low-income countries, as grants and very concessional loans. Of the remainder, however—$3 billion of technical cooperation grants, $3 billion of other grants, and $4 billion of direct loans at soft terms—a lower percentage went to help the development programmes of poorer countries in relation to their proportion of population, i.e., they had below-average receipts of ODA *per capita* in a situation in which the already staggering problems of social and economic development in poorer countries were intensified by increased prices and world recession.
>
> The amounts cited above include an estimated $6 billion of flows from OPEC countries to other developing countries, of which $2.7 billion was on concessional terms. OPEC countries' assistance made an important contribution, although their aid programmes have been highly concentrated geographically.[5]

The DAC, in 1975, provided a total net capital inflow to the LDCs amounting to $39,900 million, or 1.05 percent of the donors' total GNP.[6] In the same year, OPEC's total (concessional and non-concessional) capital flows to these LDCs was 2.9 percent of their GNP.[7]

Net disbursement of ODA by the DAC countries in 1975 was $13,600 million, or 0.36 percent of combined GNP.[8] It is difficult to determine the exact terms of OPEC's concessional aid flows and thus to distinguish between net capital flows and ODA. OECD estimates for concessional aid by OPEC in 1975 are shown in Table 10, and more recent assistance is discussed in Appendix II. There was a concessional element of about 65 percent in the ODA flows from OPEC in 1975 (Table 11); thus the grant equivalent of OPEC aid in 1975 was about 0.86 percent of GNP.[9] This proportion (with a higher concessional component than the ODA aid of DAC countries) was well above the 0.36 percent of GNP of ODA flows from DAC countries. Thus, in proportion, overall OPEC efforts were clearly greater than those of DAC countries.

This assertion is even more significant when one realizes that the GNP per capita of OPEC as a whole is *below* the world average, and significantly below that of the DAC countries.

There is, of course, considerable disparity among the OPEC donors. As expected, the richer members of OPEC—United Arab Emirates, Qatar, Kuwait, and Saudi Arabia—are obviously contributing a much larger percentage of their GNP than poorer members such as Algeria, Indonesia, and Nigeria. However, any comparision with the DAC countries indicates that OPEC members are making a *significant* contribution to the alleviation of the LDCs' balance of trade problems.

Much of OPEC's financial assistance has been in the form of multilateral aid. This has been channelled through such organizations as the OPEC Special Fund, the Arab Bank for Economic Development in Africa (ABEDA), the Special Arab Aid Fund for Africa (SAAFA), the Islamic Development Bank (IDB), the Arab Fund for Economic and Social Development (AFESD), the OAPEC Special Fund, and the Arab Monetary Fund.[10] Several of these organizations are discussed in some detail in Appendix II. Overall assistance to non-OPEC LDCs can be seen in Table 12.

Disinterested western observers would be justified in calling for more OPEC financial aid over the long-run to the non-oil producing developing countries or for more efficient matching of commitment and disbursement, promise and performance. But the past record of most DAC members could hardly be cited as a criterion for determining "proper" levels of assistance. Perhaps it would be too harsh if OPEC political leaders were to attribute some of the rhetoric frequently heard in the United States or Europe to either naiveté or hypocrisy. However, statements from high sources that ignore the

Table 10. Concessional Assistance by OPEC Members in 1975 ($ million)

	Commitments			Estimated Net Disbursements			
Donor Country	Bilateral	Multilateral	Total	Bilateral	Multilateral	Total	As % of GNP
Algeria	5.8	28.7	34.5	2.0	16.7	18.7	0.14
Iran	1,267.9	4.4	1,272.3	481.0	4.4	485.4	0.85
Iraq	294.8	25.8	320.6	190.0	34.8	224.8	1.48
Kuwait	500.0	112.7	612.7	250.0	80.5	330.5	2.75
Libya	56.3	76.1	132.4	100.0	65.7	165.7	1.62
Nigeria	20.4	9.3	29.7	20.4	9.3	29.7	0.15
Qatar	177.6	25.6	203.2	125.0	22.0	147.0	6.39
Saudi Arabia	1,438.8	237.5	1,676.3	675.0	242.2	917.2	2.64
U.A.E.	338.2	61.2	399.4	345.0	58.7	403.7	4.64
Venezuela	—	11.5	11.5	—	24.9	24.9	0.08
Total	4,099.8	592.8	4,692.6	2,188.4	559.2	2,747.6	1.35[a]

Note: [a] 1.50 percent if Nigeria is omitted.

Source: OECD, *Development Co-operation: 1976 Review* (Paris, November 1976), p. 101.

Table 11. Terms of Concessional Flows from OPEC Members, 1973–1975

	1973	1974	1975
Total Concessional Element (%)	85.0	74.0	65.0
Grants as % of Total Commitments	74.0	58.0	38.0
Grant Element of Bilateral ODA Commitments (%)	84.0	68.0	59.0
Grant Element of ODA Loans (%)	41.0	39.0	43.0
Maturity of ODA Loans (years)	17.5	14.1	15.0
Grace Period of ODA Loans (years)	5.2	4.4	4.5
Interest Rates of ODA Loans (%)	3.3	2.4	2.4

Source: OECD, *Development Co-operation: 1976 Review* (Paris, November 1976), p. 102.

fact that official DAC assistance levels are for the richest nations well *below* one-half of one percent, and that then complain about rates that are *higher* among the *nouveau riche* oil exporters (themselves still developing countries, Tables 13 and A.10), must risk being sternly censured—that is, if they are considered at all.

Others might argue that OPEC members should do even more for the LDCs, since they reap monopoly rents from their oil. However, this discussion is largely outside of the realm of pure economic theory. For instance, what is a "fair" price for a barrel of oil? Would it be "permissible" for an oil-rich nation (if OPEC did not exist) to restrict its output on the grounds that oil prices would naturally rise faster than returns on invested past oil receipts? If so, what the OPEC nations are doing by mutual agreement could conceivably happen in a "natural" fashion. What should be done with western corporations that have effective monopoly positions in export markets? How should the industrialized countries react to firms that practice reverse dumping? These questions are endless; many countries and/or their corporations can be shown to diverge widely from purely competitive practi-

Table 12. Financing of Non-Oil Developing Countries' Current Account Deficits, 1974–1976 ($ billion)

	Actual 1974		Estimate 1975		Forecast 1976	
Current Balance Deficit	–39		–49		–41	
	Financing Items					
1. Official Transfers	7.81		8.45		9.17	
(of which: Technical Co-operation)	(3.16)		(3.55)		(3.80)	
(a) DAC		5.03		5.94		6.60
(b) Multilateral Agencies		1.34		1.48		1.70
(c) OPEC		1.37		0.96		0.80
(d) Centrally Planned Economies		0.07		0.07		0.07
2. Long-Term Official and Private Capital	19.07		26.10		25.15	
ODA	5.25		6.11		6.50	
(a) DAC		2.46		3.07		3.30
(b) Multilateral Agencies		1.02		1.35		1.50
(c) OPEC		0.83		1.07		1.00
(d) Centrally Planned Economies		0.94		0.62		0.70
Other Official Flows	3.88		7.09		6.65	
(a) DAC		1.31		2.72		1.50
(b) Multilateral Agencies		1.84		2.23		3.00
(c) OPEC		0.61		1.99		2.00
(d) Centrally Planned Economies		0.12		0.15		0.15
Private (DAC)	9.94		12.90		12.00	

Table 12. Financing of Non-Oil Developing Countries' Current Account Deficits, 1974–1976 (Continued)

	Actual 1974		Estimate 1975		Forecast 1976	
Current Balance Deficit	–39		–49		–41	
			Financing Items			
(a) Direct Investment		4.50		6.00		5.50
(b) Portfolio Investment		3.73		4.60		4.00
(c) Export Credits		1.71		2.30		2.50
3. Other	9.43		10.33		8.50	
(a) IMF Oil Facility		1.15		2.43		0.00
(b) Use of Other IMF Credit		0.38		0.30		3.50
(c) Euro-currency Borrowing		7.90		7.60		5.00
4. Change in Reserve Position	+1.52		–3.53		+5.00	
5. Short-Term Capital Plus Errors and Omissions	+4.21		–0.59		+3.18	
Total 1 + 2 + 3	36.31		44.88		42.82	
Total 1 + 2 + 3 – 4	34.79		48.41		37.82	

Note: These figures are recorded on a balance of payments reporting basis. "Transfers" and "long-term capital" exclude grants for military purposes, debt relief, and reinvested earnings.

Source: OECD, *Development Co-operation: 1976 Review* (Paris, November 1976), p. 40.

ces. The uniqueness of oil arises in that it is a vital resource whose market is no longer mostly controlled by the industrialized countries or by their multinational corporations.

Several major OPEC countries, besides having a level of per capita income well below that of the industrialized countries, are themselves experiencing balance of trade problems. For example, Algeria showed a deficit in its trade account in 1974, 1975, and 1976. Furthermore, Iran, Iraq, and Nigeria could soon be facing growing balance of trade deficits. The present level of financial aid may well be a sacrifice that some OPEC members cannot bear much longer. In the future, Algeria, Iraq, Nigeria, and even Iran might not be able to go along with the richer exporters, such as Kuwait, Qatar, Saudi Arabia, and the United Arab Emirates. Given the current and expected economic problems of some of its members, overall OPEC aid might not significantly rise, in real terms, above current levels; it might even fall in real terms by the early 1980s, if the economic problems of the more heavily populated OPEC members become more prominent.

Though the oil exporters may feel some obligation to help their erstwhile compatriots in poverty among the LDCs, it does not follow that they should bear the total burden in financing the LDC trade deficits. Oil price increases, as we have seen in the previous chapter, can not be blamed for all or even most of these deficits. Furthermore, it must be recalled that none of the OPEC states are what could be termed economically developed, though all are richer now than they were in 1973, and some are much richer. Nearly all the major producers (Table 10) have been disbursing aid at, or well above, the goal (0.7 percent of GNP) set by the World Bank for official development assistance. However, with the exception of Libya and the smaller Gulf states, none of the oil exporters has yet a GNP close to the levels enjoyed by most of the industrialized countries (Tables 13 and A.10).

The 1973-1974 price increases jumped the aggregate gross product of the oil exporters by as much as $125,000 million ($375 per capita) between 1972 and 1975. Even so, the oil producers' share of the world product in 1975 was less than 4 percent compared to about 61 percent, or nearly 16 times as much, still claimed by the developed capitalist states.

On a per capita basis, heavily populated Indonesia and Nigeria kept the OPEC mean well below $1,000. The Middle East, on average, was more fortunate (Table A.10), though the range was wide. It ran from the relatively modest levels of Algeria, Iraq, and Iran (with a collective population of about 60 million) on up to the heights of the mini-states of Qatar, Kuwait, and the United Arab Emirates (with a total population of less than two million).

Again, only naiveté could explain any expectation that the massive LDC financial problems could, or even should, be solved to any great extent (say, 50 percent or more) by oil producer contributions. Yet such ideas have been voiced in official western circles, most notably during the provisional presidency of Gerald Ford.

While some OPEC members can undoubtedly do more (and, from many indications, will do more) in the future, collectively they are simply too small a part of the world economy to supply most of the funds needed by the LDCs. All of the oil producers, even the richest, are themselves developing countries with perhaps 20 to 40 years to transform their economies as they exploit a rapidly wasting resource. Thus, there must be a limit to the share of their collective incomes that they can devote to foreign economic assistance without doing serious harm either to their own development plans or to the domestic tranquility needed to carry out these vast programs. The wasting nature of the resource upon which their development depends implies that "rich," as the term is applied to the oil producers, is deceptive.[11] Rather, a distinction must be drawn between productive wealth—such as agricultural resources, capital goods, skilled labor—and non-productive wealth—such as oil in the ground. Thus Saudi Arabia or Iran, for example, can be seen trying to *achieve* real, or productive, wealth, but certainly with a considerable distance to go.

Some oil producers, like Algeria, Nigeria, and Venezuela, are already at, or very near, the limits of their ability to extend aid to the LDCs. Even if Saudi Arabia and the

Table 13. Aggregated Gross Product, 1972–1975

	1972	1974[a]		1975[b]	
		I	II	I	II
Gross World Product ($ 1,000 million)	4,120.2	5,540.1	5,555.9	6,309.2	6,330.0
Developed Countries[b]					
Gross Product ($ 1,000 millions)	2,736.2	. . . . 3,590.0		. . . . 3,851.5	
Average GNP Per Capita ($)	3,925	. . . . 5,062		. . . . 5,391	
Aggregate Gross Product as Percentage of Gross World Product	64.4	64.8	64.6	61.0	60.8
Oil Exporters[c]					
Gross Product ($ 1,000 million)	114.5	188.7	204.6	217.0	237.8
Average GNP Per Capita ($)	412	642	696	719	788
Aggregate Gross Product as Percentage of Gross World Product	2.8	3.4	3.7	3.4	3.8

Notes:

[a] The World Bank method for calculating international GNP estimates uses a multi-year average of prices and exchange rates. For example, in Tables 13 and A.10, the figures for 1972 and the first set for 1974 were computed using this method and the base periods 1970 to 1972 and 1972 to 1974, respectively. However, this method understates the income effect in the oil-producing states of the 1973 price increases. The second set of 1974 figures was calculated using only 1974 prices and exchange rates; the two sets of figures establish a range with which pre-1973 comparisons may be made. The first set of 1975 figures was based on the 1973 to 1975 period, and the second set used 1974–1975 prices and exchange rates.

[b] Developed countries are those so designated above: North America, Australia, New Zealand, Japan, South Africa, and OECD-Europe, except for Greece, Portugal, Turkey, and Yugoslavia.

[c] All OPEC members plus Bahrain, Brunei, Oman, and Trinidad-Tobago.

Source: *World Bank Atlas* (1975 and 1976 editions).

smaller Gulf states could push their largesse to 15 percent of their GNP, OPEC financial assistance would total at most (in 1975 terms) about $10,000 million a year, while recent United Nations and IMF estimates have calculated LDC needs to be at least $30,000 million a year for even modest development gains.

As long as major western nations are far from achieving the DAC goal of 0.7 percent of GNP in official assistance, it is hardly likely that OPEC leaders will feel pressured to push their own aid toward levels they feel are likely to be damaging to their own economies.

By coincidence, if official aid from the developed states had been at the recommended 0.7 percent level in 1974-1975, then the $30,000 million goal would have just about been met[12]—roughly $24,000 million from the developed countries, and close to $5,000 million from the oil exporters. However, aid from the developed states was in fact only about *half* this amount.

Not surprisingly, the non-oil LDCs continually emphasize the shortcomings of the industrialized countries in this regard, despite attempts, particularly by the United States during the Ford-Kissinger administration, to divide the LDCs from the OPEC states. Though the former have often loudly grumbled about OPEC as well as DAC aid, they still generally see the oil states as their major spokesmen—as, for example, at the 1976 preliminary sessions of the Conference on International Economic Cooperation (CIEC) when Venezuela emerged as the unofficial chairman of the LDC caucus.

In short, then, insofar as OPEC has assisted in the financing of the adjustment problems of LDCs, its responsibility for their current difficulties through higher oil prices has been somewhat blunted. However, OPEC assistance has been attacked on three related grounds: OPEC aid is geographically concentrated; it is directed towards Muslim countries; and it does not help the poorest LDCs. An OECD source made the following case, using 1974 and 1976 data:

> OPEC aid is characterized by a high degree of *geographic concentration.* [original emphasis] Both in 1974 and in 1975, 70 percent of total bilateral ODA commitments was accounted for

by only 5 recipients. Egypt and India were by far the largest recipients of OPEC assistance in 1975 and accounted together for about half of bilateral aid commitments. Syria ranked third with over 10 per cent, followed by Jordan and Oman. Pakistan, which had been the largest recipient in 1974, occupied only the sixth place in 1975. Arab countries again obtained three-fifths of total bilateral ODA commitments, as in 1974. The vast majority of developing countries did not receive any, or only small amounts of, aid until the end of 1975. In all, 35 countries obtained concessional aid commitments in 1975 (40 in 1974). At present, there is evidence of a wider geographic spread of commitments and disbursements on the part of several OPEC donors, which will undoubtedly increase the number of recipients in 1976 and possibly somewhat diminish the present high geographic concentration of aid, which is likely nevertheless to persist for some time in view of the special relationship between Arab donors and the so-called "Confrontation States."

Assistance to developing countries with *per capita* incomes of $200 or less accounted for 45 percent of OPEC bilateral ODA commitments in 1975. Although the share of the poorest countries remained practically unchanged between 1974 (46 percent) and 1975, there were some important changes inside this group of recipients. A very pronounced fall in commitments to Pakistan ($140 million, following $750 million the year before) was compensated by a sharp increase in commitments to India ($930 million as compared with $240 million). The remainder was spread over 18 recipient countries, with Indonesia, Somalia, Sudan, and Y.A.R. receiving close to $100 million each. At the other end of the range, countries with *per capita* incomes of more than $1,000 obtained only 4 percent of bilateral ODA commitments in 1975 (9 percent in 1974), almost exclusively on account of Oman. Recipients with *per capita* incomes of $200 to $374 obtained one quarter of 1975 commitments. Roughly the same share went to the group of countries between $375 and $699.[13]

Implications of religious favoritism tend to ignore several relevant considerations. Approximately a fifth of mankind professes Islam—about 700 million people, nearly all of whom live in developing countries. Many of the most distressed countries, that is, those having the severest balance of trade problems (Tables 8, 9, and A.7), are found in south and southwest Asia and in central Africa—running west from the Ganges delta to the Atlantic coast of Mauritania, north from the Congo River to the Black Sea. Historically, this region comprises the principle arena of Islamic history; not surprisingly, most of these countries have sizable Muslim populations. Whether they have been given special treatment on this account, or because they are the closest neighbors of those OPEC states that have given most of the financial aid, they are *still* deserving recipients even by the strictest economic criteria: that is, they face severe adjustment difficulties and comprise some of the poorest countries in the world.

Prior to 1973, only Kuwait (among both major Muslim and non-Muslim oil exporters) had much experience in granting and especially in administering significant amounts of development assitance. Others now find themselves with the resources to finance their own aid programs, but they have had many "start-up" problems that have caused frequent delays in implementation.

Not surprisingly, the result has been that aid has gone to areas that the donors already know best. Venezuela, for example, has directed its earliest efforts to its Caribbean neighbors. Nigerian assistance, in more modest amounts, has gone to its immediate west African compatriots. While it has been suggested that the inexperienced oil exporters could channel their contributions through international agencies, until recently similar suggestions have been for the most part ignored by western aid-giving states; even today they still prefer to maintain direct control over most of their largesse.

Finally, specific examples, particularly by 1977, belie accusations of religious or ethnic exclusivity. For example, India, non-Muslim, but a neighbor to Iran and the major Arab exporters, was the second largest recipient of OPEC aid, as is indicated in the OECD summary cited above. In fact, assistance to India might be interpreted as having been extended at least partly at the expense of Muslim Pakistan.

The Kuwait Fund for Arab Economic Development (KFAED) amended its 15-year-old charter, and in the fiscal year 1975-1976 began to grant aid to non-Arab nations. Simultaneously, it sharply increased the amount of its aid, committing almost three-fourths as much from April 1975 through early 1977 (close to $800 million) as it had from its founding in 1962 through early 1975 (about $1,100 million). About half the 1975-1977 KFAED grants went to non-Arab states in Africa and Asia. Specifically, in the fiscal year 1975-1976, KFAED made about 42 percent of its nearly $560 million in commitments (Appendix II) to non-Arab countries, with India, Pakistan, Bangladesh, Malaysia, Sri Lanka, Uganda, and Tanzania prominent among the recipients. In 1976-1977, KFAED had proposed to extend aid even farther afield—from the Philippines to several West African states.

New aid-giving institutions founded since 1974 by OPEC members, acting as individuals or in groups, have also defined their goals in broad LDC terms. Nevertheless, such organizations as the Arab Bank for Economic Development in Africa (ABEDA) and the OPEC Special Fund have had frequent problems implementing the programs that their funding allows them to support. As ABEDA president, Dr. Chedly Ayari, stated in a recent interview:

> By 1980, African investment requirements will come to three to four billion dollars . . . We have estimated potential sources for financing. The gap still left amounts to one billion dollars a year . . . ABEDA does not have the managerial capacity for such a sum, but we do for perhaps half as much.[14]

Both ABEDA and the OPEC Special Fund began substantive operations in 1976; most of their activities are not reflected in the OECD reports quoted earlier. The projects of these two agencies are listed in Appendix II, along with those of other aid-granting institutions. This post-1975 data is reasonably complete for those groups listed, but it is not as exhaustive a representation of oil-exporter assistance as the OECD figures for 1974 and 1975. Still, it further illustrates the broadening scope of OPEC assistance, both as regards the kinds of recipients and the purposes of the grants.

For example, between late 1975 and early 1977, ABEDA committed about $180 million in loans for 26 projects in 21 countries. Most loans were for 25 years (including a five-year grace period), at an average rate of interest of 3 percent. The OPEC Special Fund was founded in January 1976; it was capitalized with about $700 million during its first year of operations. Over 60 percent of this went into the oil producers' share of the International Fund for Agricultural Development (IFAD), and most of the rest was earmarked for balance of payments support to LDCs. By May 1977, commitments of nearly $200 million had been made to 49 countries for this purpose. It is envisaged that the OPEC Fund will be replenished annually by the contributing members and that future activities will include aid aimed at specific proposals. In August 1977, the Fund approved another $83 million for 18 projects in 17 countries and 4 projects of the United Nations Development Program. As can be seen in Appendix II, Special Fund recipients are located throughout the developing world.

One possible exception to this tendency to expand the base beyond the Arab or Muslim world has been the Islamic Development Bank (IDB), formed in late 1974. Despite many delays, at the time the IDB granted its first loans, it had paid-up capital of $875 million and authorization to lend up to about $2,325 million. The major backer of the IDB is Saudi Arabia, whose government leaders have been sincerely and sorely troubled by Qu-r'anic injunctions against the charging of interest; IDB transactions carry only a nominal service charge. Despite the IDB's ostensible orientation, its first loan went to Cameroon, whose population is only about 20 percent Muslim.[15]

It is obvious, however, that OPEC members should continue to spread their assistance beyond the countries they are most accustomed to dealing with, especially to southern and central Africa and to Latin America. Here the experience of KFAED can be useful, as it has been in the past, to the more recent donor countries and organizations.

The example set by KFAED should also be considered with regard to other matters. For example, between 1973 and 1975, drops were recorded in the concessional elements of OPEC loans and in the percentage of total commitments made in the form of grants, as is illustrated above in Table 11. In 1973, OPEC aid was still largely dominated by Kuwait's generous efforts, while many OPEC commitments in 1974 and 1975 were more strictly defined. There are indications (Appendix II) that the more recent ODA activity of several OPEC donors (such as ABEDA and the OPEC Special Fund) involves easier terms, which are more comparable to KFAED practices.[16] Also, unlike most industrialized countries, Kuwait (with only oil to sell) did not follow the custom of tying or restricting its aid, thus allowing recipients to shop in world markets for development implements at the best available prices, thus raising the value of such aid. Some OPEC aid granted since 1973 has been specifically targeted for oil purchases,[17] but most has been committed to projects whose foreign exchange requirements clearly involve imports from the industrialized countries. Over the next few years, however, some OPEC members may be tempted to follow American, European, and Japanese examples. Those with embryonic industrial bases, like Iran and Venezuela, could begin to see ODA as a means to boost their non-oil exports. Tying ODA to oil purchases need not reduce its net impact; however, there is general agreement that past western practices have had this effect. It seems likely that, at least over the short and medium term, industrial products originating from OPEC countries will not be very competitive in world markets,[18] and that therefore tied aid would mean reduced aid, with the degree of diminution depending both upon the goods involved and the financial terms of the ODA.[19] If the grantors of OPEC aid follow the western model in this regard, they would yield some of the advantage they currently enjoy in the political debate about LDC assistance.

The dispute over the geographical and religious concentration of OPEC aid clearly demonstrates the dilemma of non-OPEC LDCs. They obviously need more aid, yet the industrialized countries and OPEC squabble over who gives what to whom. Bangladesh, for example, cares very little about where it gets its aid, but the donors, seeking influence, want to channel their aid in such a way as to maximize their own perceived welfare. As a result, some LDCs may be relatively generously endowed by both industrialized countries and OPEC members, while other LDCs are almost totally neglected.

In the next chapter we will explore the various ways in which the non-OPEC LDCs can be helped to escape from this vicious triangle and the low-level foreign exchange trap.

Notes

1. Such solutions, in this case, include the devaluation of the peso in September 1976, which had been long delayed for reasons of domestic politics. After some fluctuation, the peso fell almost 50 percent against the dollar. In the first three quarters following devaluation, exports averaged about 125 percent higher than in the three quarters preceding this step, while average quarterly import costs climbed about 30 percent. In the first half of 1977, Mexico's trade deficit was only about 4 billion pesos (about $180 million), compared to about 17.5 billion pesos (then about $1,400 million) in the first half of 1976. Mexico, in fact, is relying on post-1973 OPEC actions to turn its chronic trade deficits around by 1980. Oil now entering world markets from its recent massive discoveries along the Isthmus of Tehuantepec will command OPEC-set prices. If Mexican oil exports hit a million barrels a day by 1979, as many observers predict, export receipts will grow by almost $5,000 million a year. Total Mexican exports in 1976, by contrast, came to less than $3,000 million. The IMF has estimated that more than 30 percent of these export earnings were needed to service Mexico's growing foreign debt. Despite all these benefits, no prudent Mexican president would lead his nation into OPEC membership. Such benefits of higher oil prices can be Mexico's *without* joining OPEC, and such an act would deny the U.S.'s largest neighbor most-favored-nation treatment, as a result of legislation adopted in 1974 by Congress in a fit of pique when apparently the only victims were thought to be Arabic-speaking.
2. Most domestic exploration has been handled by the government-owned corporation, Petrobras, whose efforts have at least in part been hampered by worldwide shortages of such vital equipment as offshore drilling rigs. On the other hand, Petrobras has been involved, with some success, in joint exploration ventures in the Middle East.
3. The severe international payments problems of a third Latin American country, Peru, were prominent in press reports by the end of 1977, but again for reasons largely based on circumstances other than world oil prices. On the one hand, domestic oil production in Peru has recently been about two-thirds of consumption (Table A.6), while the shares of the oil-exporting countries in Peru's added import bill (Table 8) and trade balance decline (Table 9) were relatively small—about 10 percent in both cases over the 1972-1975 period. Yet Peru's trade balance went from a surplus of about $150 million in 1972 to a deficit of nearly $1,200 million by 1975. The direct effects of higher oil prices—even adding an estimate of the indirect effects—played only a minor role in this decline. Spurred by the general commodity boom in the early 1970s, Peru moved into a period of greatly expanded spending, with imports more than tripling between 1972 and 1975; in large part, these were financed through massive overseas borrow-

ing. Unfortunately, while Peruvian export prices jumped 43 percent between 1971 and 1973, they plunged 54 percent in 1974, before recovering, in 1975, to 1971 levels; during the same interval, the volume of exports declined about 28 percent. Copper, fishmeal, and sugar—industries that produce half or more of Peruvian exports in most years—all underwent wide fluctuations in output and prices during this period, while the government and private sector sought (and found) credit abroad at levels in keeping with the peak earnings of the commodity boom. By late 1977, even severely slashed imports (partly due to a two-year decline in the value of the sol of about two-thirds) could not keep Peru from nearly defaulting on foreign debts, now believed to be as much as 50 percent above the 1975 level shown in Table A. 14.

4. OECD, *Development Co-operation: 1976 Review*, pp. 13-14.
5. Ibid., p. 15.
6. Ibid.
7. Ibid., p. 99.
8. Ibid., p. 16. Official development assistance is defined as assistance with a grant element of 84 percent or more; and in 1975 the actual overall grant element of total ODA commitments was 88 percent. Ibid., p. 157.
9. The director general of the OPEC Special Fund, Ibrahim Shahata, recently presented (*Middle East Economic Digest*, 9 December 1977, p. 23) even higher estimates: "In 1976, OPEC countries occupied the top six ranks among all donor countries as regards the proportion of foreign aid to GNP. While Western countries had hardly achieved in 1974-76 the disbursement-GNP ratio of 0.33 percent, OPEC countries as a group reached the 2.7 ratio, with the major donors exceeding 10 percent of their GNP. . . . The grant element in OPEC concessional flows is estimated at 81.3 percent in 1974, 73.6 percent in 1975, and 78.7 percent in 1976. In fact it is higher than such percentages indicate, considering that all OPEC aid is untied to source."
10. Several OPEC members have also participated in the efforts of such international agencies as the International Monetary Fund to ease post-1973 balance of payments problems. Borrowing through the Fund's Oil Facility were a number of developed countries who, of course, paid market interest rates. The oil-producing states, however, have also directed concessionary aid through multilateral agencies. For example, the new United Nations specialized agency, the International Fund for Agricultural Development (IFAD), began operations in 1977 with nearly $1,050 million in pledges. Of this, over $440 million came from the oil exporters, led by Iran ($126 million), Saudi Arabia ($107 million), Venezuela ($67 million), and Kuwait ($37 million). About $575 million was pledged by DAC members and $20 million by non-oil LDCs.
11. Obviously a "temporarily rich" country could exploit a wasting resource at a technologically determined maximum rate well above its own ability to absorb capital fruitfully. The surplus could then be directed by prudent investors to the most promising LDC investment opportunities. Thus, both the LDCs and the oil producers would eventually profit—provided we assume away all the difficulties linking high levels of oil production, in the short run, with thoroughly and rationally programed LDC investment schedules. It is more likely that even the most altruistic possessor of the natural resources in high demand in world markets would demand a clear picture of both its own capital absorption capacity and its investment opportunities among its neighbors.
12. This goal, however, had been established before oil prices were raised in 1973.
13. OECD, *Development Co-operation*, p. 102.
14. "L'Aide Arabe et le Financement du Developpement Africain," *Jeune Afrique*, Lettre d'Information N. 1, (January 1976); authors' translation.
15. The IDB's president, Dr. Ahmad Muhammad Ali, recently announced (*Middle East Economic Survey*, 14 November 1977) that in the first two years of the institution's activity, about $158 million had been lent for 25

projects in the 32 member states. Loans are without interest, but annual service charges of 0.75 to 3 percent are levied. In some cases, the IDB takes an equity interest in the assisted project. In 1977 borrowers included Algeria, Bangladesh, Egypt, Guinea, Jordan, Niger, Pakistan, Somalia, Sudan, Tunisia, Turkey, and Yemen People's Democratic Republic, while the IDB granted $20 million to UNESCO for several educational projects in Africa and Asia.

16. For example, OPEC Special Fund loans have carried annual service charges of either 0.5 or 0.75 percent; about half of ABEDA funding involves charges of 3 percent or less.
17. The Special Arab Aid Fund for Africa (SAAFA) was devised as a special purpose fund analogous to the International Monetary Fund's Oil Facility, to which several OPEC members were major lenders. The 1976 balance of payments support program of the OPEC Special Fund also fell into this category. SAAFA has since been merged with ABEDA, which, like current OPEC Special Fund efforts, aims its financing at specific economic development investment projects; thus, foreign exchange availability to the LDC recipient generally results in capital goods expenditures in the developed countries.
18. Some important exceptions exist; most prominent among these would be the nitrogenous fertilizers now appearing on world markets from petrochemical plants in the Gulf region.
19. There are, of course, ways for grantors to restrict ODA, other than for the specific purchase of their own exports, that may have relevance to the OPEC-LDC relationship. For example, there has already been considerable interest in joint projects to build petrochemical plants in LDCs, where the partners would be the host country and one (or more) of the oil exporters. Most of the technology needed for such plants would be purchased from industrialized countries, but these deals usually have also involved contracts for the OPEC partners to supply the petroleum inputs. On the one hand, such an arrangement may, on balance, bring the host country both investment capital and needed raw materials on generally favorable terms. On the other hand, it could turn out to give an OPEC member's national oil company protected access to a sizable and growing LDC economy—for the price of a loan granted at terms that might be scarcely distinguishable from those obtainable in commercial markets.

    Another type of restriction involves the natural resources of the recipient rather than the donor. In this regard, the Sudan serves as an excellent example. Nearly every Middle Eastern oil producer (including non-Arab Iran) has already shown interest in developing the Sudan's agricultural potential. Since the country is more or less self-sufficient in food (though at a definitely basic level), the severalfold increases envisaged for Sudanese agriculture are obviously seen by potential investors in terms of expanded exports—to other parts of a region of prevailing food deficits (cf. Askari and Cummings, *Middle East Economies in the 1970s*, ch. 3). Again the specific terms of OPEC development aid must be considered. An LDC might gain both capital at concessional terms and protected access to growing and prosperous OPEC markets, or it might find itself obligated to sell its output under unfavorable conditions in order to pay off loans whose financial terms alone earn the OPEC investor a solid return.

4. POLICY OPTIONS FOR FINANCING LDC DEFICITS AND ECONOMIC GROWTH

The general economic problem facing LDCs in the 1970s is neither new nor very different from that of the last two decades. The bottlenecks to rapid economic growth and development have been shortages of foreign exchange, savings, and skilled labor of various categories, and an inability to adapt technology to local conditions. However, the magnitude of the problems of LDCs have changed since 1973. In this chapter, we will address the need for policies to help them overcome their short- and long-run economic problems.

To finance their trade deficits, the non-OPEC LDCs have been forced to rely on additional borrowing. The magnitude of this borrowing since 1969 is shown in Table 14. The IMF has summarized the situation as follows:

> The external public debt outstanding of 75 non-oil exporting developing countries at the end of 1976 is estimated by the World Bank at $171.1 billion, of which $122.8 billion had been disbursed by the end of 1976. By the end of 1973, the last full year before the increase in oil prices and the severe 1974-75 recession, the external public debt of these countries was $93.0 billion, of which $66.8 billion had been disbursed by the end of the year. At the end of 1975, the comparable figures were $140.8 billion and $100.0 billion. At the end of 1976, the public debt outstanding of 9 oil-exporting developing countries was $41.5 billion, or nearly $8.5 billion higher than at the end of 1975. Thus, at the end of 1976, total public debt outstanding of all 84 countries regularly surveyed by the World Bank was $212.6 billion.
>
> Of the $22.8 billion increase in public debt disbursed during 1976 to the 75 non-oil exporting developing countries, $10.5 billion was from

Table 14. External Public Debt of 84 LDCs, by Region, 1970–1975
(Debt outstanding, disbursed only, in $ million as of December 31)

	1970	1973	1974	1975
		Official Plus Private		
Total–Regions Without Oil				
Africa South of Sahara	4,310.4	6,776.0	8,649.3	10,583.8
East Asia and Pacific	3,837.7	6,493.1	7,985.7	10,660.8
Latin America and Caribbean	15,005.1	24,416.0	32,152.3	39,476.9
North Africa and Middle East	3,017.7	4,143.7	4,684.3	7,298.0
South Asia	11,933.9	16,443.2	18,509.5	19,922.2
More Advanced Mediterranean Countries	5,687.1	8,501.9	10,350.6	12,100.9
Total–75 Non-Oil Countries	43,791.8	66,773.8	82,331.8	100,042.7
Total–9 Oil Exporting Countries	7,475.3	15,864.2	17,873.0	20,939.7
Total–84 Countries	51,267.1	82,638.1	100,204.8	120,982.4
		Total Official		
Total–Regions Without Oil				
Africa South of Sahara	2,795.4	4,543.7	5,795.5	7,027.5
East Asia and Pacific	1,839.7	3,913.0	4,673.8	5,858.4
Latin America and Caribbean	7,828.0	10,737.2	13,074.8	14,952.8
North Africa and Middle East	2,447.3	3,433.4	3,734.9	6,064.2
South Asia	11,245.9	15,691.8	17,692.6	19,156.2
More Advanced Mediterranean Countries	3,894.4	5,723.4	6,286.1	6,714.5
Total–75 Non-Oil Countries	30,050.6	44,042.5	51,257.6	59,773.6
Total–9 Oil Exporting Countries	5,275.7	9,433.6	10,668.0	11,522.9
Total–84 Countries	35,326.3	53,476.1	61.925.6	71,296.5

Table 14. External Public Debt of 84 LDCs, by Region, 1970–1975 (Continued)

	1970	1973	1974	1975
			Total Private	
Total-Regions Without Oil				
Africa South of Sahara	1,515.0	2,232.3	2,853.9	3,556.2
East Asia and Pacific	1,998.0	2,580.0	3,311.9	4,802.4
Latin America and Caribbean	7,177.1	13,678.8	19,077.5	24,524.1
North Africa and Middle East	570.4	710.3	949.4	1,233.8
South Asia	688.0	751.4	816.9	766.1
More Advanced Mediterranean Countries	1,792.7	2,778.5	4,064.5	5,386.4
Total–75 Non–Oil Countries	13,741.2	22,731.3	31,074.2	40,269.0
Total–9 Oil Exporting Countries	2,199.6	6,430.6	7,205.0	9,416.9
Total–84 Countries	15,940.8	29,161.9	38,279.2	49,685.9

Source: *IMF Survey* (5 September 1977), p. 282.

> official sources and $12.3 billion was from private sources. In 1975, official sources had accounted for $8.5 billion of the $17.8 billion increase. The rise in the proportion accounted for by private sources reflected the fact that from the early 1970s, the group of middle-income countries tended to borrow increasingly from them.[1]

Much of the expansion in private financing for the middle-income countries has been in the Eurodollar market. The World Bank estimated total Eurodollar flows to LDCs at $1,400 million in 1971, $3,600 million in 1972, and $9,100 million in 1973. Morgan Guaranty Trust Company of New York put Eurodollar loans to non-OPEC developing nations at about $8,000 million in 1975, increasing to $11,000 million in 1976. If such financing were available and if there were no problems associated with the growing Eurodollar debt, then we could recommend that the market continue to accommodate the financing needs of developing nations. This is not, however, the case.

Antonio Ortiz Mena, President of the Inter-American Development Bank, indicated some of these problems in a 1974 speech:

> . . . the Eurocurrency market has provided a large volume of financing for the region [Latin America] in the last two years, but . . . this financing is being obtained on conditions that, without careful planning, can frustrate orderly management of the external debt and even weaken the internal savings efforts of our countries.
>
> As you know, the usual form of loans in the Eurocurrency market is the revolving credit with a fluctuating interest rate. Although the credit is extended for periods that have been lengthening gradually to 10 and 12 years—and 14 and 15 years in some cases—in practice the credit is renewed every six months, each time at the interest rate prevailing in the London market (interbank offer rate, IBOR). Since 1969 there have been sharp fluctuations from a low of slightly over 5 percent to a high of 11 percent . . . It should be noted that the loans usually are amortized in full at the end of the agreed period and that the resources are completely united.

> These operations are transacted with scant knowledge of the feasibility of the projects, since brokers are commonly used to promote lending operations, especially in the developing countries. Obviously, such practices can lead to the excessive use of credit and to an improper allocation of financial resources . . . This observation is even more to the point if it is kept in mind that the countries sometimes resort to the Eurocurrency market to finance the total cost of an investment.
>
> . . . in actual figures the Eurocurrency market supplied resources to those countries [eight major Latin American countries in 1973] for more than double the financing authorized by the international agencies [the Inter-American and World Banks].
>
> Finally, we note that the oil crisis is forcing the industrialized countries into the Eurocurrency market in order to finance their balance of payments deficits, which could displace the developing countries . . .
>
> The foregoing considerations suggest that advisability of broadening the Bank's activities so as to increase its advisory services . . .[2]

Such financing might be appropriate for some developing nations whose debt service to export ratios are *low*, but it would not be for the many nations whose ratios are high and which in addition have serious internal and external economic problems. In Table 15, the relevant ratios for debt and debt service for non-oil developing countries are shown. While LDC debts climbed sharply from 1973 to 1976, debt service was still equal to about 10 percent of exports. However, within the group of LDCs, there is considerable variation as to how much of export earnings go to pay for past borrowings, as can be seen from Table A.11. If we compare this list with what was discussed before in chapter 2,[3] we see that by 1975 a number of countries with relatively high debt service to export earnings ratios were also among those whose trade balances were most seriously affected by the direct effects of higher oil prices. These include (Table 16) Uruguay, Chile, Brazil, Greece, and Turkey. The first two in particular have experienced very sharp increases in their debt

Table 15. Ratios of Public Debt to GNP and Exports and of Debt Service to Exports of Non-Oil LDCs, 1973–1975 (percent)[a]

Ratio of:	1973	1974	1975	1976[b]
Debt to GNP	11.9	11.9	13.4	15.0
Debt to Exports	76.5	71.1	85.0	88.9
Debt Service to Exports	10.1	8.4	9.9	10.0

Notes:

[a] Outstanding, disbursed only.

[b] Estimated.

Source: *IMF Survey* (5 September 1977), p. 283.

service to export ratios between 1973 and 1975. On the other hand, the list in Table 16 also contains several countries that are either minor oil exporters or self-sufficient in this regard (Argentina, Bolivia, Burma, Colombia, Egypt, Mexico, and Zaire) and some countries whose ratios improved during this two-year period (India, Korea, and Nicaragua). Others on the list might be termed chronic high debt countries, whose export earnings were heavily entailed with debts even before higher oil prices. Thus, again what we see is a mixed picture; by no means can the overall debt problems of the LDCs be traced primarily to OPEC pricing policies.

Among those countries whose debt service to exports ratios remained constant or actually dropped between 1973 and 1975 are a number of countries that, as we noted in chapter 3, experienced sharp increases in export earnings during this period. Included here are several countries (Korea, India, Jordan, Singapore, and Taiwan) that enjoy expanded markets among the oil producers themselves. On the other hand, it seems more likely that the overall decline among LDCs in this ratio is due to an easing of debt service terms during this period—payment moratoria and easier credit for refinancing older debts. In the long-run, their problems could become worse—if, for

Table 16. LDCs with Higher than Average External Public Debt Service to Exports Ratios, 1973–1975

Country	Debt Service as Percent of Exports	
	1973	1975
LDC Average	10.1	9.9
Uruguay	22.9	45.9
Chile	10.9	28.4
Mexico	24.3	25.9
Egypt	34.1	25.7
Peru	23.6	23.8
Argentina	17.6	22.0
Sudan	12.2	21.3
Sri Lanka	12.9	20.3
Pakistan	16.9	19.1
Burma	26.7	18.0
Bangladesh	2.2	17.9
Zaire	8.5	16.5
India	19.9	16.3
Bolivia	15.5	15.2
Brazil	13.2	15.0
Central African Empire	4.0	12.3
Colombia	13.6	11.8
Greece	10.1	11.6
Korea	12.2	11.6
Turkey	13.4	11.5
Nicaragua	17.4	11.3
Costa Rica	10.3	10.7
Uganda	6.8	10.5

Source: See Table A.11.

example, easier terms in the immediate post-1973 period are followed in the late 1970s and early 1980s by higher interest rates or even a general shrinking of credit availability.

The latter circumstance could result in either of two eventualities, one less catastrophic (and probably more likely) than the other, given the growing role of private lenders in the LDC credit search (Table 17). On the one hand, debt service costs can continue to climb at about the 20 percent average annual rate of growth that prevailed between 1970 and 1975, pushing up the debt service to export ratio above its 1975 level of 9.9 percent. More and more LDCs would become poor credit risks as far as private lenders are concerned. Since credit to the LDCs from the private sector grew nearly twice as fast as from official lenders between 1970 and 1975 (28 versus 17 percent annual average rate), and actually accounted for about three-fifths of the net increase in LDC debt by 1975, lack of access to this part of the market would severely constrain LDCs with continuing trade deficits and/or the need to refinance older debts. Considering the future service costs of already contracted debt (Table 18), restricted access to private credit markets seems a likely eventuality for many LDCs.

The second possible scenario involves default by one or more major LDC borrowers on debts contracted in the private market. Morgan Guaranty Trust estimated that by late 1976 the non-oil exporting LDCs owed private banks almost $77,000 million; of this total, about two-thirds, or $52,000 million, was owed to U.S. banks. Another New York banking firm, Salomon Brothers, pointed out the recent lure of foreign lending. In 1973, they estimated that the 13 largest U.S. banks earned $477 million overseas, about a third of their total earnings; by 1975, this figure had climbed to $836 million, almost half their total earnings. For several banks (most notably Bankers Trust, Chase Manhattan, and Chemical Bank) domestic earnings actually declined between 1970 and 1975; the best that could be said for most of the rest was that they experienced sluggish domestic growth during this period. All, however, had dramatically more rewarding international dealings.[4]

Unfortunately, it is not very difficult to imagine the consequences of serious default by any one of half a dozen of the most troubled LDC borrowers. A recent report prepared for the U.S. Senate Foreign Relations Committee's subcommittee on foreign economic policy indicated that about two-thirds of all lending to LDCs by U.S. banks has been done by only six institutions. Such a degree of concentration among creditors clearly indicates that a major default could do far more than merely dry up the private credit available to the most overextended LDCs. A major bank failure in the U.S. or western Europe would certainly affect the domestic economies of all the industrialized countries significantly.

Table 17. Debt and Debt Service Costs of 84 LDCs, 1970–1975 ($ 1,000 millions)

	1970	1973	1974	1975
Total Debt Outstanding	71.2	117.1	148.0	174.2
Official Creditors	50.3	79.3	95.7	106.8
Private Creditors	20.9	37.8	52.3	67.5
Disbursed Debt Only	51.3	82.7	100.3	121.2
Official Creditors	35.4	53.5	62.0	70.9
Private Creditors	15.9	29.2	38.2	50.2
Debt Service	5.9	11.0	12.9	14.3
Amortization	4.1	7.8	8.6	8.8
Interest Payments	1.8	3.2	4.3	5.5

Source: *IMF Survey* (6 June 1977).

Table 18. Projected Service Costs for Existing Debt of 84 LDCs, 1977–1985 ($ millions)

Type of Creditor	Outstanding Debt (including undisbursed) 31.12.1975	Debt Service Costs (projected)			
		1977	1978	1980	1985
Bilateral Official	70,291	5,439	5,495	5,400	3,998
Multilateral Official	37,020	2,094	2,402	2,842	2,796
Banks	40,558	7,491	8,468	7,439	1,299
Other Private	26,044	4,591	4,340	3,345	1,213

Source: *IMF Survey* (5 September 1977).

Until now, however, the availability of private credit has in part made it possible for many LDCs to experience fairly good growth; in fact, they have generally done better recently than the DAC countries (Table 19). Still, the potential problems associated with continued heavy reliance on private capital markets makes it imperative for us to consider other options.

One alternative policy would be to extend the remedy of outright grants indefinitely, perhaps in the form of some type of rebate on LDC oil purchases that would be geared to pre-1973 prices and consumption levels, and contain an annual escalator for consumption levels. For the short-run, or even longer for some countries, this avenue might offer considerable promise of easing trade balance problems. But even in a limited time frame, such measures attack only part of the problem. They would do nothing for the non-oil part of the deficit and would help primarily those LDCs with the largest oil imports, doing little for others. These "others" include many of the poorest countries—those with the lowest per capita energy use—and some of the largest (for example, Colombia, Egypt, Mexico, and Zaire) that supply all or most of their own oil but still have huge trade deficits.

Over the long-run this approach offers no real solution even to those who would reap immediate gains. It amounts, in fact, to "pensioning off" the recipients of the aid. It might even give them disincentives to develop alternative long-run energy-use policies, in a way analogous to what many observers have pointed out as an unwanted side effect of U.S. food assistance in the 1960s—increasing, rather than decreasing, dependence on foreign sources. Also, since the oil rebates, or subsidies, would be tied in some way to 1973 consumption, such a program would, in itself, do little to promote a degree of economic development (so closely related to energy use) rising very far above 1973 levels. Furthermore, while the deficits of the smallest LDCs might be covered by grants indefinitely, those of several of the more populous LDCs are already too large for such aid to be expedient for much beyond the end of the 1970s.

If the origins of the current trade problems of the LDCs are so complex, there is little to be gained from trying to assess blame. If blame is to be assessed at all, there is enough for all parties—the OECD countries, the oil exporters, and, in many cases, the LDCs themselves, which have a history of haphazard economic policies and programs.

Rather, concern should be with policies designed to halt and reverse the deteriorating trends of the last de-

Table 19. GNP and GNP Per Capita Growth Rates

		Annual Average % Change		
	Population 1974 (million)	Real GNP Growth 1965–1973	Real GNP Per Capita Growth 1965–1973	Real GNP Per Capita Growth 1975[a]
Non–Oil Developing Countries:				
Higher–Income Group (above $375 per capita)	509	7.2	4.8	0.0
Middle–Income Group (between $200 and $374 per capita)	178	5.2	2.4	1.6
Lower–Income Group (below $200 per capita in 1973)	1,040	3.6	1.2	1.6[b]
(of which India)	(586)	(3.8)	(1.5)	(2.9)
DAC Countries	650	4.6	3.6	–2.0

Notes:

[a] Change over previous year. Secretariat estimate based on preliminary World Bank data.

[b] Excluding Laos, Cambodia, and Vietnam, for which no data are available.

Source: OECD, *Development Co-operation: 1976 Review* (Paris, November 1976), p. 44.

cade; it is from this concern that the North-South dialogue has evolved. The Sixth and Seventh Special Sessions of the United Nations General Assembly in 1974 and 1975 were fora for much of this dialogue. The General Assembly then established the Conference on International Economic Cooperation (CIEC), a group of 27 countries—made up of developed states, oil exporters, and LDCs—whose assignment was to begin detailed negotiations. An agenda containing many possibilities for easing the LDC trade dilemma was proposed for their consideration. The major emphases were in four areas: creating and maintaining buffer stocks of major raw materials; linking the prices of major LDC exports to those of their imports from the developed states; easing tariff and nontariff barriers that retard the efforts of LDCs to capture more of the value-added of final consumer goods by processing raw materials before export; and increasing the levels and quality of financial and technical assistance to the LDCs.

CIEC held a series of preliminary meetings in 1976, dividing itself into several committees to prepare for the negotiations called for in the United Nations. The 1975 Special Session of the General Assembly ended optimistically, signalled by a willingness on the part of the developed states, especially the United States, to give the major points of the proposed agenda serious consideration. But both the 1976 preparatory meetings and the formal CIEC sessions, ending in June 1977, failed to make much progress on two key items: indexing LDC import prices and initiating an integrated buffer-stock program for major raw materials. The following ten commodities are on the list of raw materials most discussed: cocoa, copper, cotton, hard fibers, iron ore, jute, rubber, sugar, tea, and tin. Developed states have voiced opposition to the General Assembly proposal on several fronts. They fear that its cost will be prohibitive and that even a partial suspension of "normal" market activities for these commodities will be severely inflationary. In what many LDCs fear may be a "divide-and-conquer" strategy, the U.S. and other developed states have suggested a step-by-step approach, implying that some basic commodities might rate buffer-stock support while others would not.

Even the scheduling arrangements for the formal CIEC sessions deepened LDC apprehensions about the developed countries' intentions. These sessions were originally supposed to be held in December 1976. But only a few days before the scheduled opening, a delay of several months was announced, ostensibly because of the pending change of administrations in the United States; however at least two other factors were also important in the postponement. First was the lack of consensus in the preliminary sessions and second was the desire of many industrialized countries (and perhaps some of the oil exporters) to know the results of an OPEC meeting, also scheduled for December 1976, before embarking on CIEC negotiations. Thus, this delay was to some extent part of a diplomatic game between the OECD and OPEC states, with the non-oil LDCs as almost voiceless and increasingly desperate onlookers.

If meaningful policies are to be designed to halt the deterioration in the LDCs international financial positions, then the North-South dialogue must turn from infighting to serious negotiation over the points raised in the General Assembly special sessions. Negotiation does not need to be limited to these points only. These sessions were, for example, dominated by the LDCs which were in the majority, and the resulting agenda may well contain items whose adoption in the form suggested by the General Assembly is either politically improbable or economically unwise, or both. Sympathetic western observers of the LDCs have pointed out that *both* practical politics and practical economics—the former often termed the art of the possible, the latter the science of scarcity—must be present on all sides of any negotiations that hope to lead to a new international economic order. These two vital constraints—political possibility within the framework of independent nation states and limited resource availability on "Spaceship Earth"—have, perhaps, never before been (*together*) so much the center of attention both in the industrialized and in the developing states as they have been in the post-oil crisis period.

On the one hand, the desire of the LDCs to capture more of the eventual value-added for their own economies by processing raw materials before export, even to the point of exporting final products, is not only understandable but quite economically defensible, especially for final products that are labor intensive. Political reality, however, requires that the LDCs remember that western industrialized countries are mostly parliamentary democracies, and no matter how much sense there may be in the argument, for example, that Brazil should export shoes rather than leather and rubber to the U.S., the voices of domestic business and labor interests are almost always going to be heard more clearly in the U.S. Congress than dispassionate arguments for international economic rationality. Thus, the lowering of western barriers to LDC exports in the manufactures categories must be a gradual process, as in fact it has been for the easing of trade restrictions among the industrialized countries themselves since World War II. Such caution is all the more advisable in light of the neo-protectionist wave that has almost engulfed intra-OECD trade relations as a result of the mid-1970s recession.

On the other hand, sincere and serious economic questions have been raised about most of the major points on the CIEC agenda. Will the linking of the prices of raw materials and manufactured goods through some indexing mechanism mean institutionalizing the worldwide inflation of the early 1970s? What sort of capital will be demanded to support buffer stocks of so many commodities? An open-ended commitment could lead perhaps to $10 to $15 thousand million or more—money that, if directed to LDC aid, might be used better in productive investment rather than in inventory maintenance.

The long-run implications of some of these programs for the LDCs have, for example, been found wanting by economists sympathetic to their plight, such as Harry Johnson, who pointed out that the buffer-stock proposals for agricultural commodities can actually reduce average producer incomes when demand instability is present, or increase income variability when supply instability is found. Thus, in practice the twin goals of higher and more stable producer income are unlikely to be achieved.[5]

If LDC requirements of relatively stable (around a growth trend line) foreign exchange earnings cannot, at least for agricultural commodities, be achieved through the mechanism of buffer stocks, then the LDCs must accept the "unwisdom" of such a policy, while the developed countries must move toward finding some other means to achieve the same end. It is clear that rational development planning is nearly impossible in the face of wide and unpredictable secular fluctuations in foreign exchange earnings. If institutionalized market manipulations cannot solve this problem, then perhaps countercyclical credit mechanisms (supported by capital from the industrialized and oil-exporting states) tied to commodity price movements can help. What emerges from a consideration of these questions is the conclusion that unwise or immoderate LDC demands can now only be met by imaginative western responses, showing a willingness at least to consider policies that stray from the not very successful neo-classical economic paths of the past.

One possible structure that bears some promise for a new economic order is trilateralism. The technology of the industrialized countries and the financial assistance of both industrialized and richer oil-exporting states would be combined to provide for productive investment in, not merely subsistence aid to, the LDCs.

If such investment is to ease current trade problems, it must involve production of goods now figuring on both the export and import sides of LDC ledgers. For exports there is the problem of expanding those sectors of LDC economies that can partially or completely process the commodities still generally exported as raw materials. The developed states remain the major customers of the LDCs; they are the ones that must face the often domestically unpopular decisions that would allow LDC semi-processed goods and light manufactures to have wider access to their markets.

But the oil exporters also have responsibilities in this situation. Their purchases (Tables 5 and 6) from the LDCs have grown rapidly in the last few years, though absolute levels remain quite low. This small absolute volume of OPEC imports from LDCs is to be expected in light of the similarity of products produced in the LDCs in general, that is, including OPEC. Their non-oil exports are in large part competitive with one another. Current OPEC development programs could favor LDC exports or they could result in highly protective commercial policies that would cut short the export success several LDCs have enjoyed since 1972 (Tables 6 and 7).

The expansion of LDC imports since the late 1960s has been rapid on three fronts—energy, of course, then capital goods, and food. For the first two categories, the non-oil producing LDCs must remain dependent mostly on outside sources for at least the next several years. As indicated above, some LDCs have the potential to become self-sufficient in energy, but most have no evidence to support seriously any such hopes. Many of the smaller LDCs could be subsidized or pensioned off—a policy that is feasible, perhaps necessary, for the next several years. But even for the least populous countries this would be no real solution over the long run. Energy imports (and other kinds of purchases such as capital goods) must be financed primarily through more economically conventional foreign exchange receipts, that is, from exports or from new investment.

For food imports, on the other hand, the situation is different. LDC cereal imports particularly have grown much faster than trade in general, experiencing a tenfold quantitative increase in the last 20 to 25 years (Table A.12). With the only major cereals surplus area in the developing world, Latin America, becoming less important in this regard, the major gainers have been the wheat exporters among the developed countries—the United States, Canada, and Australia.

Cereals imports cost the developing regions an average of about $4,100 million a year in the 1970-1972 period. Since then, imports have continued to climb steadily; for example, fragmentary figures from the Food and Agricultural Organization (FAO) indicate tonnage increases of about 80 percent for wheat and about 60 percent for maize from 1972 to 1976.[6] These figures do not include one major importer in the past, India, which enjoyed large harvests in 1975 and 1976, a major factor in the relatively good trade balances it experienced during those years. This favorable circumstance is generally attributed to the luck of the monsoon, rather than any basic increase in agricultural productivity, and a reversal in the generosity of nature during the late 1970s would mean a serious escalation in these import figures.

Despite the high cost of these imports, they should be seen in the context of overall cereals production in the LDCs. It is apparent (Table A.12) that production has also grown rapidly, nearly doubling from 1950 to the early 1970s. Demand has grown faster primarily because of population pressures and to a lesser extent because of rising incomes. Overall, there has been at least some improvement in average nutritional standards. The United Nations World Food Conference reported that, contrary to popular belief, production in the developing countries grew just about as fast as it did in the developed countries in the 1950s and 1960s—at an average annual rate of about 2.9 percent. The problem of the

LDCs was that they had an average rate of growth for demand of about 3.5 percent, as opposed to about 2.5 percent in the developed countries. The World Food Conference found that food production grew at an annual average rate of 4 percent or better from 1952 to 1972 in only 25 of 96 LDCs. The average growth rate of domestic demand for food outstripped that of food production in only 5 of the developed countries, but in 54 of the 86 LDCs for which both rates could be estimated.[7]

Thus, economic assistance policies that aim to keep the gap between output and demand from widening must stress *both* population control and major improvements in agricultural productivity. Again, trilateral approaches offer promise—that is, western agricultural technology working together with investment capital from both the developed and oil-exporting states in the agricultural resources of the LDCs. The oil states have, in fact, a special stake in the success of such investments. On the one hand, the application of many of the techniques of modern agriculture in the LDCs will involve considerable petroleum inputs (fertilizers, small tractors, irrigation pumps, and so forth). On the other hand, the oil exporters, particularly those in the Middle East, are themselves very much a food-deficit region. Since some of them have rather scanty prospects for enhancing their own agricultural output to any considerable degree, neighbors endowed with better soils and more abundant water offer attractive investment opportunities—as, for example, Saudi Arabia's interest in Sudanese agriculture.

While the creation of a worldwide reserve of food grains is high on the lists of priorities of both the World Food Conference and the General Assembly special sessions, reserves are not the solution to the LDC food deficits: the reserves are for emergencies, the deficits are an annual event. The long-standing U.S. food aid program probably prevented famines on many occasions, but it might also be partly responsible for the failure, over the long run, of supply to keep up with demand in many LDCs. As was said above, the trade imbalance problems of the LDCs cannot be solved by subsidies, by "pensioning off" the neediest countries; this is true for both oil and food. The resolution adopted at the close of the

General Assembly Seventh Special Session presents the situation succintly: "The solution to world food problems lies primarily in increasing rapidly food production in the developing countries."

Another form of long-term development assistance receiving attention in the early stages of the North-South dialogue is to link, in some way, the creation of new Special Drawing Rights (SDRs) to the capital needs of the LDCs. This is not a new proposal; indeed its roots go as far back as Keynes' original proposal of the Bancor as an international currency unit to help war-ravaged Europe pay for its own reconstruction. More recently, Park stated the problems facing adoption of such a system as a means of aiding the LDCs:

> The case for a link derives fundamentally from an analogy between official reserves and private money balances in a national monetary system. In addition to the primary effect of increasing liquidity, issuance of national money is also associated with a transfer of resources to the government, or with the financing of real expenditure within the country through the lending activities of the banking system. As Professor Tinbergen (Netherlands National Committee, 1972) argues, the creation of new money always implies that the first recipient gets the money without having produced something, and this privilege should be given to the poor countries of the world community rather than to the rich. Opponents of the link argue contrarily that the liquidity-creating and liquidity-distributing functions of a banking system are conceptually distinct and that the primary emphasis in the creation of SDRs should be placed on the liquidity-creating function (Group of Ten, 1965). This latter view prevailed in 1969 when the Special Drawing Account was established in the IMF, and, as a result, the link was not incorporated in the present SDR system.[8]

This form of assistance is what may be needed for developing nations to insure substantial economic growth and to ease the burden of debt financing, which in the past has been handled through multinational agencies and private institutions, and which now involves the growing possibility of default.

A rather different proposal is also connected with reformation of the international financial structure and the place of the SDR within it. Modigliani and Askari suggested a modified SDR with constant purchasing power and defined in terms of a basket of internationally traded commodities.[9] Such an asset would be much more attractive to countries with considerable financial surpluses than any other currently available reserve asset, all of which are subject to the withering effects of inflation. Oil exporters like Saudi Arabia or the Gulf states, despite their huge increases in imports, have continued to amass holdings of both official and unofficial reserves. Government holdings have been particularly subject to inflation, and the argument has been made that these countries would be better off cutting production and leaving oil in the ground until it is needed to purchase goods. Saudi Arabia, for example, could cover its current level of imports by producing only about 5 million barrels a day, only a bit more than half the level of output in 1977. Furthermore, if, as some observers have noted,[10] many of the purchases and projects already undertaken by the oil exporters involve a considerable degree of waste, at least some of the exporters might take steps to curb this waste after the first flush of the oil revenue boom wears off. They too would then be susceptible to arguments in favor of cutting back production and slowing down their rush toward economic transformation.

However, if there were a reserve asset largely free from inflationary loss of purchasing power, any move to restrict production for financial reasons might be counteracted. The creation of an SDR so modified should be a major goal for OPEC. Political leaders in some oil countries have argued instead for indexing of oil prices, which OPEC could do unilaterally.[11] Such "reform" would not eliminate problems related to the general inflation of the industrialized countries, though it would moderate the effect of relying on a currency strongly subject to inflation (such as pound sterling in 1974 through 1977, or the dollar in 1977 and early 1978). On the other hand, a constant purchasing power asset would allow OPEC nations to develop more systematically rather than wastefully and haphazardly. Furthermore, if this asset were a revised SDR that would be available through the agency of the IMF for lending to the LDCs, then a major new source of official development funding would be created. Of course, such credit would not be enough by itself to finance LDC needs. But with conventional assistance from the industrialized and oil-exporting states, together with other steps that might emerge from the CIEC negotiations, it would make the future development prospects of the LDCs much brighter.

Up to this point, the roles of OPEC and the industrialized countries have dominated our considerations of policies to help the LDCs in their balance of trade difficulties and the design of successful development programs. However, there are also policies that LDCs can pursue on their own, managing their own economies in a more efficient way. This general statement covers all areas of policy relative to reduction of LDC waste. For example, the cost of military expenditures is one case in point:

> The compounded effect of lost GDP [to the LDCs] in 1973 due to resources diverted for military purposes from 1953 to 1973 was calculated to be roughly $102.6 billion in constant 1970 U.S. dollars; in 1973 dollars this was equivalent to a loss of nearly 26 percent of aggregate GNP of all developing countries in 1973. Even with an extremely conservative estimate, the loss would be around 12 percent of aggregate GNP. This loss is an average for all developing countries. Individual LDCs were affected much more; in some cases, such as Jordan, the loss was more than the country's entire GDP for 1973.[12]

In addition, the maintenance of modern military equipment requires a great deal of skilled labor input, sometimes the scarcest input of all. For example, the operation of a squadron of 24 F-4 phantom jets requires between 694 and 996 people, with 80 percent being skilled technicians. The estimates of the GNP loss cited above are conservative since they do not incorporate the full implication of this loss of skilled manpower.

The most prominent policy to correct a basic payments deficit is devaluation. Such a policy is appropriate when the Marshall-Lerner condition holds, that is, when the sum of the elasticity of demand for exports and imports is greater than one. However, in the case of LDCs one may expect that the demand for both its exports and its imports may be inelastic. More importantly, if some LDCs devalue their currencies, others will follow suit, and in the end none of them may benefit. More likely, their collective export revenues will go down. Thus, under the present situation, the fact that competitive devaluations have not yet occurred is to be applauded.[13]

It is not our intention here to determine a specific package of solutions to the financial situation of the LDCs. This is very much a situation of individual prescriptions for individual problems. Still, several basic points should be emphasized. In the first place, the origins of the development problems and of the trade deficits of the LDCs are more complicated than merely today's higher oil prices. It should be clear that OPEC, for all its faults, has become a scapegoat for the politicians and journalists of the OECD countries. LDCs would *still* be in very serious trouble even if the western nations had been able to "scare" OPEC into immediate dissolution in early 1974. Secondly, many of the developing countries have problems so difficult to resolve, and are economically so disadvantaged relative to both the industrial countries and the leading members of OPEC, that debt financing at or near prevailing commercial rates cannot solve their long-term needs but can only add to their problems. Indeed, debt service difficulties could lead to massive defaults that would in turn severely damage the economies of both the industrialized and oil-exporting states.

Finally, the OECD and OPEC states must cooperate, rather than blame each other, and work with the LDCs in some trilateral fashion. Yet, the LDCs receive short shift in more than just the surface circumstances of the OECD-OPEC conflict so often portrayed in the world press since late 1973. Even solutions to this conflict which are less outrageously uncaring of LDC problems than the recently reported "secret" U.S.-Saudi Arabia pact (which allegedly would channel a major share of Saudi surpluses into the financing of U.S. budget deficits for the indefinite future)[14] could still augur poorly for the LDCs. The utter callousness of these alleged U.S.-Saudi Arabian agreements (disregarding not only LDC problems, but also those of the other OECD members) might explain the incredulity that greeted publication of their details. Still, many observers fear that a brief period of OECD-OPEC confrontation might prompt the co-opting of the latter into the "economic establishment," and that OPEC prosperity would quickly erase memories of past poverty and recent spokesmanship for LDC problems.

Though this assessment may be unduly pessimistic, it may still be true that the best chance for pleading the LDC cause comes while the traditional economic balance of power remains disturbed. Since 1973, both OECD and OPEC have had to bid for LDC support, while all three groupings are at least paying lip service to the concepts of the resource limitations of "Spaceship Earth." Within each bloc there are forces emerging that have the potential to further the cause of long-run cooperation. For example, because of the suspicion that the U.S. might arrange an alliance with the major oil producers that would not be in their interest, both the European Economic Community and Japan have often been unwilling to accept the U.S. lead in the recent North-South dialogue. Within OPEC, some countries (for example, Algeria, Indonesia, Iraq, Libya, and Nigeria) are likely to remain closely tied to the LDC cause over the long run, because of economic necessity and/or ideological conviction, no matter how prosperous (and forgetful) others might become.

This sort of diversity within and among the blocs will surely determine the framework within which any long-term assistance programs emerge from fora such as the CIEC. The roles played by various countries will be governed not only by their actual, but also by their potential, wealth. Thus major (but heavily populated) oil producers like Indonesia, Mexico, and Nigeria should not be seen as donors, but as recipients, of international assistance; responsibility for LDCs like Colombia, Egypt, Syria, or Zaire, which are self-sufficient in oil *and* have serious trade balance problems, cannot be disowned by either OECD or OPEC.

We have by and large emphasized the recent problems of the developing countries. However, both the developed countries and several OPEC members have their own domestic economic problems. Many of the developed countries have been burdened with high unemployment and high inflation simultaneously. At the same time real economic growth has also been below average. (Such a combination certainly explains, at least in the short-run, why a U.S. administration would consider the kind of alliance with Saudi Arabia mentioned above, or why, if the opportunity should arise, Japan, Federal Germany, France, or Britain would also push for such a deal).

OPEC countries, on the other hand, have had severe difficulties in embarking on meaningful programs of long-run economic development. Numerous bottlenecks (shortages of skilled labor, crowded port facilities, inadequate transportation and communication networks, overburdened electric generating plants, problems in formulating rational economic policies, and so on) have brought about a tremendous waste of economic resources. Thus, facing their own particular problems, both developed countries and OPEC members may be less than anxious to undertake wholehearted efforts to help the LDCs.

With each passing day the problems of the LDCs become more critical; the melodrama of such a statement should not detract from its urgency. Surely the time has come for OECD and OPEC to accept the fact that the fundamental problems facing the LDCs cannot be blamed on one side of the Mediterranean or the other. Even if some sage could allocate blame accurately, how would such an exercise help the LDCs?

What will this relationship be?—trilateral cooperation or a vicious triangle? Continued recrimination is the luxury of the rich and the newly rich; the LDCs can only be losers.

Notes

1. *IMF Survey* (5 September 1977), p. 282.
2. Antonio Ortiz Mena, address at the Inaugural Session of the Fifteenth Meeting of the Board of Governors, Santiago, Chile, 1 April 1974, as distributed by the Inter-American Development Bank. Actual quote taken

from Carlos F. Diaz-Alejandro, *Less Developed Countries and the Post-1971 International Financial System,* Essays in International Finance No. 108.

3. See above, pp. 24-27.
4. Judith Miller, "Sounding Alarms on Foreign Debt," *New York Times* (18 September 1977), section 3, p. 1.
5. Harry G. Johnson, "Commodities: Less Developed Countries' Demands and Developed Countries' Response," (Paper delivered to the Massachusetts Institute of Technology Workshop on Specific Proposals and Desirable DC Response to LDC Demands Regarding the New International Economic Order, Cambridge, Massachusetts, 17-20 May 1976).

   The problem has also recently been discussed by World Bank Staff economists Ezriel M. Brook, Enzo R. Grilli, and Jean Waelbroek in *World Bank Staff Working Paper No. 262,* "Commodity Price Stabilization and the Developing Countries: The Problem of Choice," a shorter version of which appeared in *Finance and Development* (March 1977). They pointed out that, while both producers and consumers experience a welfare gain from price stabilization, the income effect will be positive for producers and negative for consumers if the instabilities are predominantly on the supply side. On the other hand, demand instabilities would presage a reversal of effects after a stabilization program. They then examined a group of 17 primary commodities for evidence of situations where LDCs would clearly gain—that is, positive gains from *both* the welfare and income effects. Such were indicated for six agricultural goods produced by the LDCs—cocoa, coffee, wool, jute, cotton, and sugar—but statistical significance was lacking for the last two. For the two commodities of which LDCs were net importers—wheat and rice—the income effect was found to be positive, but statistically significant only for wheat. Estimates indicate income effect gains for the developed countries in the case of rice, maize, and rubber, and inconclusive results for tea and sisal. Similarly, positive income effects seem to be indicated for the developed countries for four mineral products—copper, zinc, lead, and bauxite—while for tin the estimates were again inconclusive.
6. FAO, *Monthly Bulletin* (July/August 1976 and February 1977).
7. United Nations World Food Conference, *Assessment of the World Food Situation: Present and Future* (1974), pp. 3, 51-54.
8. Y. S. Park, *The Link Between Special Drawing Rights and Development Finance,* Essays in International Finance No. 100.
9. Franco Modigliani and Hossein Askari, *The Reform of the International Payments System,* Princeton University International Finance Section, Essays in International Finance No. 89; the application of this proposal to the problem of oil exporter financial surpluses was discussed in some detail by Askari and Cummings, *Middle East Economies in the 1970s,* pp. 499-503.
10. Including the authors, *Middle East Economies in the 1970s,* p. 493 fn.
11. This resulted in the further spreading of inflation through a multiplier effect. One recent argument for such a link, "The New World Order: The Energy Dimension," by Reza Fallah and Fereidun Fesharaki, *OPEC Review* (October 1977), proposed that a link to the export prices of the major industrialized countries, rather than to the import prices of the oil producers, would both preserve OPEC purchasing power and avoid undue harm to non-oil LDCs. While the first point might be true, it is hard to see how the LDCs can avoid being hurt when they must pay higher prices for *both* oil and the exports of the industrialized countries.
12. Hossein Askari and Michael Glover, *Military Expenditures and the Level of Economic Development,* pp. 59-60.
13. Empirically even this may be questioned. See Hossein Askari, Yves Bizien, and Ekram Hossain, "A Study of the Effectiveness of Devaluations and Revaluations in the Post-War Era," *Economia Internazionale,* Vol. XXVI, August-November, 1973.

14. These highly controversial charges were first made by the respected British bimonthly, *International Currency Review (ICR),* as well as in its sister journal, *Middle East Currency Reports* (May 1977). The charges have been emphatically denied by both Saudi Arabia and U.S. officials, but had received surprisingly little comment or substantive full refutation in major western publications by early 1978. The *ICR* framework involves a pledge of U.S. military support against internal and external opponents of the Saudi regime in return for long-term Saudi oil production at levels consistent with demand in the developed countries, Saudi moderation in OPEC price deliberations, and, most importantly, annual Saudi purchases of $10-25 billion in special non-marketable and long-lived U.S. treasury notes. The latter feature would allow U.S. administrations to continue indefinitely the sizable budget deficits of recent years without exhausting the private capital market or having much effect on domestic interest rates. Cynics might explain such a deal in terms of the current Saudi regime swapping oil for its medium-term security, though the oil might perhaps best be left in the ground for future Saudi generations, while receiving paper that could conceivably be renounced by a future U.S. administration unwilling to continue with hundreds of billions of dollars of foreign debt and playing on Arabophic sentiments to avoid the problem. Alternatively, they might suggest that the Ford and Carter administrations were willing to postpone political and fiscal problems by amassing something which the U.S. has previously avoided—a sizable foreign debt, which (growing at $10-25 billion a year in current prices) could prove to be a major share of total U.S. debt within a generation. If the cynics avoid an exclusively Saudi-U.S. framework, they could also talk of OPEC billions that at least in part could be invested in LDCs, or of U.S. preoccupation with the preservation of certain politically conservative oil-rich regimes and continued disinterest in basic LDC problems.

APPENDIX I
SUPPLEMENTARY TRADE-RELATED STATISTICS

Herein are included more detailed data, often for larger samples than could be conveniently included above in the text.

Table A.1. Selected Economic Parameters for Sample of LDCs

Country	Population (1,000's)	GNP Per Capita	Imports Per Capita	Exports Per Capita	Energy Use Per Capita (Kilograms of Coal Equivalent)
India	581,911	$ 120	$ 3.84	$ 4.20	201
Brazil	101,051	760	47.33	39.62	646
Bangladesh	74,000	80	4.81	1.21	31
Pakistan	66,230	120	10.06	10.26	188
Mexico	56,047	890	52.41	33.28	1,269
Philippines	40,219	280	33.93	26.57	309
Thailand	39,400	270	37.68	27.44	300
Turkey	37,930	600	39.86	23.53	628
E. Afric. Comm.	37,283	150	26.16	22.24	102
South Korea	32,905	400	76.64	49.36	961
Burma	29,509	80	4.60	4.16	56
Ethiopia	26,550	90	7.13	6.28	31
Argentina	24,282	1,640	78.43	79.94	1,861
Zaire	23,438	140	24.50	26.79	76
Yugoslavia	20,956	1,010	153.65	106.77	1,883
Sudan	17,051	130	19.86	21.07	125
Afghanistan	16,634	90	10.16	7.49	67
Morocco	15,903	320	48.98	40.25	257
Cent. Amer. Common Market	15,572	460	69.46	67.39	348
Taiwan	15,424	660	163.63	188.98	1,400[a]
Peru	14,531	620	54.89	65.09	650
Sri Lanka	13,180	120	25.40	24.53	140
Nepal	12,020	90	6.24	1.93	12
Malaysia	11,302	570	141.62	152.04	556
Chile	10,230	720	91.99	83.62	1,361
Ghana	9,313	300	32.14	46.12	184
Portugal	8,994	1,410	243.03	143.17	1,026
Greece	8,930	1,870	262.65	97.49	2,048
Madagascar	8,301	150	24.38	19.77	71

Table A.1. Selected Economic Parameters for Sample of LDCs (Continued)

Country	Population (1,000's)	GNP Per Capita	Imports Per Capita	Exports Per Capita	Energy Use Per Capita (Kilograms of Coal Equivalent)
Mozambique	8,283	380	39.50	21.12	141
Cameroon	6,206	250	48.86	37.02	100
Y.A.R.	6,200	100	12.99	0.70	30
Ivory Coast	5,887	380	75.93	92.68	370
Upper Volta	5,714	70	10.36	3.58	14
Mali	5,370	70	14.62	7.77	24
Guinea	5,243	110	13.70	8.72	94
Malawi	4,833	110	26.83	16.81	56
Zambia	4,646	430	121.70	163.26	557
Haiti	4,454	130	14.78	9.50	31
Dominican Rep.	4,432	520	86.26	78.38	224
Niger	4,357	100	15.68	12.43	31
Hong Kong	4,160	1,430	930.94	834.04	1,232
Senegal	4,070	280	68.33	52.82	169
Rwanda	3,980	70	8.62	4.85	13
Chad	3,870	80	27.60	20.41	13
Burundi	3,580	80	8.74	7.26	17
Israel	3,210	3,010	770.09	358.01	2,914
Somalia	3,042	80	25.50	14.27	40
Uruguay	2,995	950	62.33	71.48	900
Lebanon	2,977	940	288.75	118.33	1,073
Benin	2,947	110	31.67	12.38	42
Sierra Leone	2,787	160	42.91	42.55	123
Papua New Guinea	2,596	410	112.31	84.13	250
Jordan	2,540	340	106.63	18.75	388
Paraguay	2,416	410	28.91	35.67	173
Singapore	2,185	1,830	1,552.40	1,000.96	2,060
Togo	2,119	180	40.06	23.72	70
Jamaica	1,967	990	315.91	192.68	1,439
Cent. Afric. Emp.	1,710	160	19.92	23.26	57
Panama	1,570	920	280.54	77.27	846
Y.P.D.R.	1,560	110	68.86	30.36	360
Liberia	1,452	310	123.07	183.81	432
Mauritania	1,257	200	53.25	102.14	112
Lesotho	1,165	100	4.46[a]	0[b]	N.A.
Mauritius	860	410	139.74	123.43	239
Namibia	800	N.A.	3.75[b]	66.25[b]	N.A.
Guyana	772	410	189.12	183.94	931
Botswana	641	230	12.95[b]	31.36	N.A.
Cyprus	635	1,460	495.85	210.69	1,419
Fiji	551	650	286.97	137.79	487
Guinea-Bissau	510	330	62.75	4.90	41
Gambia	493	130	50.55	39.15	73
Reunion	470	1,210	390.87	110.43	484
Swaziland	463	330	22.46[a]	144.92[a]	N.A.

Table A.1. Selected Economic Parameters for Sample of LDCs (Continued)

Country	Population (1,000's)	GNP Per Capita	Imports Per Capita	Exports Per Capita	Energy Use Per Capita (Kilograms of Coal Equivalent)
Surinam	398	870	362.41	428.82	2,834
Martinique	343	1,330	504.26	150.82	1,077
Guadeloupe	342	1,050	433.48	116.43	550
Malta	320	1,060	549.81	211.16	1,156
Equatorial Guinea	313	260	99.04	83.07	87
Cape Verde	280	340	85.71	7.14	90
Comoros	260	170	56.15	11.15	47
Macao	260	270	397.27	274.54	263
Barbados	239	1,000	592.30	187.48	1,175
Solomons	179	280	55.87	50.28	223
Western Samoa	153	250	124.18	32.68	126
Belize	132	660	338.74	191.08	547
Fr. Polynesia	130	2,680	746.15	338.46	908
New Caledonia	126	5,010	1,656.11	1,506.51	10,247
Djibouti	101	1,580	465.35	69.31	505
Tonga	95	210	76.84	30.53	N.A.
New Hebrides	92	480	281.52	144.57	622
Sao Tome	77	470	90.91	77.92	104
Gilbert and Ellice Is.	58	360	68.97	155.17	192
Bermuda	55	4,710	2,501.81	563.45	3,239
Seychelles	55	370	290.91	36.36	317
Fr. Guiana	52	1,360	869.96	88.88	853
Nauru	6.5	5,000[a]	923.08	3,061.54	7,116

Notes: Population, U.N. mid-year estimates 1973; GNP Per Capita, 1973; Imports and Exports Per Capita, 1972; Energy Use Per Capita, 1974.

N.A. indicates data not available.

[a] Estimated.

[b] Excluding South Africa.

Sources: *U.N. Statistical Yearbook; World Bank Atlas.*

Table A.2. Shifts in the Trade Balances of the Largest LDCs, 1972–1974 and 1972–1975

Country	1972–1974			1972–1975		
	Absolute Change			Absolute Change		
	All Countries (1)	Oil Exporters (2)	(2) as % of (1)	All Countries (1)	Oil Exporters (2)	(2) as % of (1)
India	– 1,380.0	– 702.5	50.9	– 2,041.6	– 411.5	20.2
Brazil	– 5,437.0	– 1,928.0	35.5	– 4,145.0	– 1,785.0	43.1
Bangladesh	– 481.5	– 74.6	15.5	– 854.6	– 66.2	7.7
Pakistan	– 637.2	+ 16.8	N.A.	– 1,090.8	– 205.0	18.8
Mexico	– 2,061.2	– 124.7	6.1	– 2,754.6	– 13.0	0.5
Philippines	– 474.4	– 468.8	98.8	– 1,133.2	– 486.2	42.9
Thailand	– 276.0	– 251.4	91.9	– 578.5	– 382.2	66.1
Turkey	– 1,625.4	– 638.1	33.1	– 2,717.5	– 566.8	20.9
E. Afric. Comm.	– 628.6	– 278.0	44.2	– 472.4	– 291.3	61.7
S. Korea	– 1,496.3	– 624.2	41.7	– 1,295.5	– 781.9	60.3
Burma	+ 31.2	+ 24.6	N.A.	+ 4.2	+ 27.0	N.A.
Ethiopia	+ 41.0	– 8.5	N.A.	– 68.7	– 13.3	19.4
Argentina	+ 258.9	– 180.9	N.A.	– 1,035.0	– 31.1	3.0
Zaire	+ 311.0	– 45.1	N.A.	– 143.9	– 31.9	21.7
Yugoslavia	– 2,756.7	– 370.9	13.5	– 2,642.0	– 272.4	10.3
Sudan	– 172.2	+ 8.8	N.A.	– 573.3	– 81.5	14.2
Afghanistan	– 64.4	– 28.6	44.4	– 77.9	+ 6.9	N.A.
Morocco	– 68.0	– 127.7	187.8	– 885.9	– 118.6	13.3
Cent. Amer. Common Market	– 720.6	– 185.3	25.7	– 612.3	– 257.2	42.0
Taiwan	– 1,854.3	– 625.3	33.7	– 1,014.2	– 320.5	31.6

Table A.2. Shifts in the Trade Balances of the Largest LDCs, 1972–1974 and 1972–1975 (Continued)

Country	1972–1974			1972–1975		
	Absolute Change			Absolute Change		
	All Countries (1)	Oil Exporters (2)	(2) as % of (1)	All Countries (1)	Oil Exporters (2)	(2) as % of (1)
Peru	– 158.6	– 76.2	48.0	– 1,032.6	– 139.8	13.5
Sri Lanka	– 140.6	+ 9.0	N.A.	– 182.1	– 9.0	4.9
Nepal	– 1.6	+ 0.1	N.A.	– 19.6	+ 0.8	N.A.
Malaysia	– 38.0	– 174.8	460.0	+ 161.8	– 154.0	N.A.
Chile	+ 655.1	– 114.8	N.A.	+ 89.0	– 68.2	N.A.
Total LDCs (103)	–28,375.1	–10,330.3	36.4	–35,666.8	– 9,009.5	25.2

Notes: N.A. indicates not applicable.

The percentage of the absolute change in the overall trade balance of a given LDC due to the change in its balance with the oil-exporting states is of interest here only if both numbers are negative. The countries listed are those with more than ten million people.

Source: *Direction of Trade* (Annual Editions, 1975 and 1976).

Table A.3. Percentage Changes in Imports of the Largest LDCs, 1972–1974 and 1972–1975

Country	1972–1974			1972–1975		
	From World	From Oil Exporters	Percent of Total Increase Directly Related to Oil Exporters	From World	From Oil Exporters	Percent of Total Increase Directly Related to Oil Exporters
India	126.6%	419.7%	37.4%	177.3%	412.7%	26.2%
Brazil	196.2	467.8	27.1	184.2	474.0	29.2
Bangladesh	207.3	a	12.6	258.0	a	10.3
Pakistan	161.1	401.7	20.1	219.5	699.6	25.7
Mexico	121.8	236.7	4.1	124.1	102.6	1.7
Philippines	152.3	341.1	23.7	171.3	389.6	24.0
Thailand	111.9	517.5	28.1	114.9	496.6	26.3
Turkey	149.8	580.0	25.7	213.4	653.0	20.3
E. Afric. Comm.	91.9	325.5	29.4	83.7	348.5	34.6
S. Korea	171.8	290.6	16.7	188.4	430.9	22.6
Burma	42.7	700.0	14.5	25.6	–100.0	– 3.5
Ethiopia	51.3	212.6	26.0	72.9	263.0	22.7
Argentina	90.0	780.8	19.8	107.9	428.3	9.1
Zaire	105.2	601.2	8.3	94.0	438.6	6.7
Yugoslavia	134.3	811.2	11.2	139.0	811.2	10.8
Sudan	96.4	300.0	3.4	205.2	2,686.5	14.3
Afghanistan	55.4	353.6	31.7	68.7	44.0	3.2
Morocco	145.2	752.9	13.9	229.6	703.4	8.1
Cent. Amer. Common Market	126.5	266.2	14.6	125.0	365.9	18.3

Table A.3. Percentage Changes in Imports of the Largest LDCs, 1972-1974 and 1972-1975 (Continued)

	1972-1974			1972-1975		
Country	From World	From Oil Exporters	Percent of Total Increase Directly Related to Oil Exporters	From World	From Oil Exporters	Percent of Total Increase Directly Related to Oil Exporters
Taiwan	176.7	340.4	17.1	134.4	278.5	18.4
Peru	92.0	343.1	12.2	204.4	585.8	9.3
Sri Lanka	96.1	a	6.0	104.1	a	22.0
Nepal	16.4	0	0	44.9	0	0
Malaysia	159.7	182.9	9.3	120.3	183.8	12.3
Chile	103.1	163.2	12.8	76.1	153.1	16.3
Total LDCs (103)	125.7%	356.3%	18.9%	133.3%	370.6%	18.5%

Note: [a] 1972 base is zero.

Source: *Direction of Trade* (Annual Editions, 1975 and 1976).

Table A.4. Percentage Changes in Exports of the Largest LDCs, 1972–1974 and 1972–1975

	1972–1974			1972–1975		
Country	To World	To Oil Exporters	Percentage of Total Increase Directly Related to Oil Exporters	To World	To Oil Exporters	Percentage of Total Increase Directly Related to Oil Exporters
India	59.3%	259.4%	24.5%	78.6%	458.6%	32.7%
Brazil	98.6	800.0	13.0	116.5	1,079.7	14.8
Bangladesh	32.0	206.7	10.6	24.7	658.1	43.7
Pakistan	64.3	297.9	53.0	54.8	218.4	45.6
Mexico	89.9	44.3	1.2	56.4	143.9	6.2
Philippines	150.1	483.7	1.5	112.7	1,553.1	6.3
Thailand	128.1	414.4	15.6	104.2	127.1	5.9
Turkey	71.7	72.7	50.5	57.0	198.9	17.4
E. Afric. Comm.	42.0	– 32.9	N.A.	41.9	– 19.8	N.A.
S. Korea	174.6	231.4	4.2	212.9	675.3	8.5
Burma	72.4	1,619.0	38.2	31.5	1,185.7	64.2
Ethiopia	102.5	163.5	12.3	41.6	173.1	25.9
Argentina	102.5	810.1	8.1	52.5	786.4	15.3
Zaire	145.7	139.3	0.4	63.0	139.3	1.0
Yugoslavia	70.1	222.6	7.2	82.0	416.3	11.5
Sudan	31.4	144.2	17.6	22.3	129.7	22.3
Afghanistan	23.5	109.1	4.1	30.7	972.7	28.0
Morocco	166.0	114.2	2.6	141.0	109.6	2.9
Cent. Amer. Common Market	56.9	223.1	2.4	70.5	270.8	2.4

Table A.4. Percentage Changes in Exports of the Largest LDCs, 1972–1974 and 1972–1975 (Continued)

	1972–1974			1972–1975		
Country	To World	To Oil Exporters	Percentage of Total Increase Directly Related to Oil Exporters	To World	To Oil Exporters	Percentage of Total Increase Directly Related to Oil Exporters
Taiwan	89.4	108.1	6.6	81.5	193.7	12.9
Peru	60.8	120.4	2.3	31.8	115.7	4.2
Sri Lanka	57.1	a	15.3	52.6	a	39.7
Nepal	46.1	a	0.9	60.8	a	5.7
Malaysia	146.5	140.5	2.5	121.5	190.7	4.0
Chile	190.0	324.1	0.6	94.2	1,665.5	6.0
Total LDCs (103)	98.6%	212.4%	7.6%	91.8%	332.8%	12.5%

Notes: [a] 1972 base is zero.

N.A. indicates not applicable.

Source: *Direction of Trade* (Annual Editions, 1975 and 1976).

Table A.5. Percentage of Trade of Largest LDCs with Oil Exporters, 1972–1975

	Percent of All Imports from Oil Exporters		Percent of All Imports to Oil Exporters	
Country	1972	1975	1972	1975
India	11.3	20.9	5.6	17.6
Brazil	10.8	22.9	1.6	8.7
Bangladesh	0.0	7.4	3.8	10.0
Pakistan	8.1	20.2	11.4	23.5
Mexico	2.1	1.9	2.4	3.8
Philippines	10.6	19.1	0.5	3.6
Thailand	6.1	16.9	4.8	5.4
Turkey	6.5	15.9	4.8	9.5
E. Afric. Comm.	8.3	20.2	4.9	2.8
S. Korea	9.9	18.2	2.7	6.6
Burma	0.9	0.0	1.7	16.7
Ethiopia	6.3	13.2	6.2	12.0
Argentina	2.3	5.8	1.0	6.0
Zaire	1.4	4.0	0.4	0.7
Yugoslavia	1.8	7.1	2.1	6.5
Sudan	1.1	10.0	3.8	7.2
Afghanistan	4.9	4.2	0.9	7.3
Morocco	2.6	6.4	3.7	3.3
Cent. Amer. Common Market	6.9	14.4	0.6	1.3
Taiwan	8.9	14.4	5.4	8.8
Peru	3.3	7.3	1.1	1.9
Sri Lanka	0.0	11.2	0.0	13.7
Nepal	0.0	0.0	0.0	2.1
Malaysia	8.1	10.4	2.6	3.4
Chile	8.1	11.6	0.3	3.1
Total LDCs (103)	6.6	13.4	3.4	7.8

Source: *Direction of Trade* (Annual Editions, 1975 and 1976).

Table A.6. Oil Production among the Non-OPEC LDCs, 1972–1975

Country	Production (Average, 1000's bbl/d)	Consumption (Average, 1000's bbl/d)	Production–Consumption Ratio
India	153	499	30.7%
Brazil	172	729	23.6%
Mexico	541	635	85.2%
Turkey	64	233	27.5%
Burma	20	25	80.0%
Argentina	413	478	86.4%
Zaire[a]	25	20-25	100-120%
Yugoslavia	69	235	29.4%
Cent. Amer.[b] (Guatemala)	6-10	65-75	8-15.5%
Peru	71	115	62.0%
Malaysia	100	115	87.0%
Chile	30	130	23.1%

Notes: 1972-1975 averages unless otherwise indicated.

[a] Production by early 1976 of several fields that began commercial exploitation as early as 1973. At this level or higher, Zaire would no longer belong to the non-oil LDC category.

[b] Commercial production in Guatemala began in 1976. This output level is estimated for fields ready for exploitation in 1977. Consumption is for all five members of the Central American Common Market.

Source: *International Petroleum Encyclopaedia, 1976; Petroleum Economist* (various issues).

Table A.7. Trade Balance–Export Ratios, LDCs, 1972-1975

Country	1972	1975	Country	1972	1975
India	+ 8.5%	– 42.0%	Lebanon	–142.6%	– 77.4%
Brazil	– 19.5	– 56.8	Benin	–155.8	–259.9
Bangladesh	–299.1	–290.0	Sierra Leone	– 0.8	– 38.3
Pakistan	+ 1.9	–102.6	Papua New Guinea	– 33.5	+ 0.1
Mexico	– 57.5	–125.7	Jordan	–468.8	–414.0
Philippines	– 27.7	– 62.9	Paraguay	– 7.1	– 1.5
Thailand	– 37.3	– 44.5	Singapore	– 55.1	– 51.3
Turkey	– 69.4	–238.2	Togo	– 68.9	– 38.2
E. Afric. Comm.	– 18.2	– 52.9	Jamaica	– 63.9	– 43.4
S. Korea	– 55.3	– 43.2	Cent. Afric. Emp.	– 14.3	– 43.0
Burma	– 10.4	– 5.3	Panama	–263.1	–213.2
Ethiopia	– 13.5	– 38.6	Y.P.D.R.	–126.8	+ 39.3
Argentina	+ 1.9	– 33.7	Liberia	+ 33.0	+ 16.1
Zaire	+ 8.5	– 8.8	Mauritania	+ 47.9	– 9.1
Yugoslavia	– 43.9	– 89.0	Lesotho[a]	N.A.	–173.7
Sudan	– 5.8	–135.2	Mauritius	– 13.2	– 11.4
Afghanistan	– 35.8	– 75.2	Guyana	– 2.8	+ 5.7
Morocco	– 56.9	– 66.4	Botswana[a]	+ 58.7	+ 80.6
Cent. Amer.			Cyprus	–136.8	–104.0
Comm. Market	– 3.1	– 36.0	Fiji	–108.3	– 57.2
Taiwan	+ 13.4	– 11.8	Guinea-Bissau	–1,180.0	–518.0
Peru	+ 15.7	– 94.7	Gambia	– 29.1	– 20.3
Sri Lanka	– 3.5	– 38.4	Reunion	–254.0	–565.3
Nepal	–223.3	–191.4	Swaziland[a]	+ 84.5	+ 75.9
Malaysia	+ 6.8	+ 7.3	Surinam	+ 15.5	+ 1.2
Chile	– 10.0	+ 0.1	Martinique	–234.4	–238.3
Ghana	+ 30.3	+ 2.0	Guadeloupe	–272.3	–259.1
Portugal	– 69.7	– 97.4	Malta	–160.6	–127.4
Greece	–169.4	–131.9	Equatorial		
Madagascar	– 23.3	+ 19.6	Guinea	– 19.2	+ 25.0
Mozambique	– 87.0	– 84.4	Cape Verde	–1,100.0	–2,550.0
Cameroon	– 32.0	– 33.7	Comoros	–403.4	–377.1
Y.A.R.	–1,728.4	–2,596.3	Macao	– 44.7	– 15.7
Ivory Coast	– 18.1	+ 5.2	Barbados	–215.9	–102.8
Upper Volta	–189.2	–294.6	Solomon Is.	– 11.1	–216.7
Mali	– 88.2	–463.8	Western Samoa	–280.0	–393.1
Guinea	– 57.2	– 22.5	Belize	– 77.3	– 1.0
Malawi	– 59.6	– 82.8	French Polynesia	–120.5	–1,045.8
Zambia	– 25.5	– 14.4	New Caledonia	– 9.9	– 22.7
Haiti	– 55.7	– 75.5	Djibouti	–571.4	–215.2
Dominican Rep.	– 10.0	+ 14.0	Tonga	–151.7	–210.2
Niger	– 26.1	–118.3	New Hebrides	– 94.3	–232.7
Hong Kong	– 11.6	– 12.3	Sao Tome	– 16.7	– 28.6
Senegal	– 29.4	– 55.4	Gilbert and		
Rwanda	– 77.7	–127.3	Ellice Is.	+ 55.6	0
Chad	– 71.4	–140.6	Bermuda	–344.0	–404.4
Burundi	– 20.4	– 93.4	Seychelles	–1,400.0	–400.0

Table A.7. Trade Balance–Export Ratios, LDCs, 1972–1975 (Continued)

Country	1972	1975	Country	1972	1975
Israel	–155.1	–209.1	Fr. Guiana	–878.8	–3,500.0
Somalia	– 78.6	–124.6	Nauru	+ 69.8	+ 72.8
Uruguay	– 12.8	– 45.7			
			All LDCs (103)	– 28.9%	– 56.9%

Notes: Positive sign indicates trade surplus, negative sign a deficit.

N.A. indicates data not available.

[a] Omitting trade with South Africa.

Source: *Direction of Trade* (Annual Editions, 1975 and 1976).

Table A.8. Trade of Selected Minor Oil Exporters, 1972-1975 ($ millions)

		Imports		
Country	Trade with	1972	1974	1975
Bolivia	World	$ 173.0	$ 390.0	$ 557.9
	Developed Countries	116.2	238.2	340.4
	Oil Exporters	0.2	0.1	0.3
Colombia	World	859.0	1,597.2	1,494.8
	Developed Countries	744.4	1,333.5	1,277.9
	Oil Exporters	7.2	17.6	18.9
Egypt	World	888.8	2,351.5	3,534.0
	Developed Countries	390.0	1,598.5	2,442.5
	Oil Exporters	21.6	28.0	42.7
Syria	World	539.2	1,229.7	1,685.5
	Developed Countries	303.3	741.0	998.5
	Oil Exporters	25.4	57.8	62.3
Tunisia	World	459.6	1,135.6	1,421.8
	Developed Countries	380.9	868.8	1,142.2
	Oil Exporters	17.0	90.2	89.1

Source: *Direction of Trade* (Annual Editions, 1975 and 1976).

Table A.8. Trade of Selected Minor Oil Exporters, 1972–1975 (Continued)

Exports			Balance of Trade		
1972	1974	1975	1972	1974	1975
$ 240.4	$ 650.5	$ 519.5	+ 67.4	+ 260.5	– 38.4
169.3	400.3	302.5	+ 53.1	+ 162.1	– 37.9
0.0	0.0	0.2	– 0.2	– 0.1	– 0.1
865.9	1,422.0	1,470.2	– 6.9	– 175.2	– 24.6
673.5	1,024.8	1,098.5	– 70.9	– 308.7	– 179.4
16.8	43.2	91.7	+ 9.6	+ 25.6	+ 72.8
825.2	1,517.1	1,576.3	– 63.6	– 834.4	–1,957.7
194.8	507.1	483.0	–196.1	–1,091.4	–2,987.5
32.2	68.3	100.8	+ 10.6	+ 40.3	+ 58.1
287.3	784.3	930.0	–251.9	– 445.4	– 755.0
87.4	263.7	453.2	–215.9	– 447.3	– 545.3
16.6	33.9	53.9	– 8.8	– 23.9	– 8.4
310.9	925.9	857.5	–148.7	– 209.7	– 564.3
243.8	586.9	517.9	–137.1	– 281.9	– 624.3
19.3	82.8	94.9	+ 2.3	– 7.4	+ 5.8

Table A.9. Trade Balance—Export Ratios, Selected Minor Oil Exporters, 1972-1975

Country	1972	1974	1975
Bolivia	+28.0%	+40.1%	– 7.3%
Colombia	– 0.8	–12.3	– 1.7
Egypt	– 7.7	–55.0	–124.2
Syria	–87.7	–56.8	–81.2
Tunisia	–47.8	–22.7	–65.8

Note: Positive sign indicates trade surplus, negative sign a deficit.

Source: *Direction of Trade* (Annual Editions, 1975 and 1976).

Table A.10. GNP Per Capita, Selected Oil-Producing and Developed Countries, 1972-1975

	1972	1974[a]		1975[a]	
		(I)	(II)	(I)	(II)
All Oil Producers	$ 412	$ 642	$ 696	$ 719	$ 788
Middle East/ North Africa Oil Producers	990	1,610	1,749	1,810	2,010
Algeria	550	640	770	780	830
Iran	760	1,050	1,330	1,440	1,560
Iraq	600	970	1,200	1,280	1,380
Kuwait	10,560	7,790	11,730	11,510	12,060
Libya	3,300	4,030	5,070	5,080	5,050
Qatar	5,060	5,930	9,140	8,320	9,090
Saudi Arabia	1,190	2,040	2,940	3,010	3,810
U.A.E.	9,770	8,340	13,470	10,480	12,530
Other Oil Producers					
Indonesia	120	150	180	180	200
Nigeria	190	250	310	310	340
Venezuela	1,520	1,800	2,180	2,220	2,260
Developed Countries	3,925	5,062		5,391	
Australia	3,970	5,330		5,640	
Canada	4,910	6,190		6,650	
France	4,080	5,440		5,760	
German Federal Republic	4,810	6,260		6,610	
Italy	2,210	2,820		2,940	
Japan	3,140	4,070		4,460	
Spain	1,520	2,490		2,700	
Sweden	5,510	7,240		7,880	
United Kingdom	2,740	3,590		3,840	
United States	5,590	6,670		7,060	

Note: [a] See Table 13, note a.

Source: *World Bank Atlas* (1975 and 1976 Editions).

Table A.11. External Public Debt and Debt Service Payments of 84 LDCs
($ millions, as of 31 December 1975)

Country	Total Debt	Of which: Official	Banks	Other Private	Debt Actually Disbursed	Debt Service as Percent of Exports[a]: 1973	1974	1975
Africa South of the Sahara								
Benin	$ 161.2	$ 135.0	$ 13.5	$ 12.7	$ 93.7	2.3%	5.2%	4.7%
Botswana	184.5	184.5	—	—	144.2	2.8	2.5	4.5
Burundi	34.7	25.8	2.0	6.8	14.6	3.0	3.3	5.7
Cameroon	688.1	603.4	73.1	11.6	361.7	4.7	4.2	5.3
Cent. Afric. Emp.	109.5	98.3	2.4	8.8	63.4	4.0	8.1	12.3
Chad	163.3	144.6	7.4	11.3	69.5	3.4	3.1	5.7
Congo	616.0	393.6	1.6	220.9	336.4	6.6	4.1	9.4
E. Afric. Comm.	344.5	197.9	—	46.6	297.8	—	—	—
Ethiopia	673.7	644.0	26.8	2.9	381.3	6.5	5.7	7.7
Gambia	22.1	22.1	—	—	13.2	1.1	0.8	0.6
Ghana	797.3	613.3	—	184.1	603.8	2.2	2.4	3.2
Ivory Coast	1,537.4	697.8	373.0	466.7	973.8	7.2	8.3	9.8
Kenya	1,071.2	949.6	26.7	95.0	520.4	5.3[b]	4.0[b]	4.0[b]
Lesotho	22.8	22.1	0.4	0.4	13.2	3.3	2.1	2.8
Liberia	274.6	256.1	6.9	11.6	169.7	5.3	4.8	5.2
Madagascar	236.5	225.6	3.9	6.9	167.9	5.2	3.3	2.9
Malawi	330.3	312.2	5.1	12.9	240.8	9.1	9.1	7.4
Mali	471.0	457.0	0.8	13.1	327.0	1.6	2.8	3.1
Mauritania	410.2	348.6	9.6	51.9	154.6	2.0	3.8	15.3
Mauritius	124.8	123.8	—	1.0	43.1	1.7	1.0	1.0
Niger	131.7	126.0	3.3	2.3	105.9	2.8	5.4	6.0
Rwanda	82.5	80.0	—	2.4	21.2	0.3	1.3	1.2
Senegal	503.0	365.7	95.6	41.8	288.6	8.3	5.5	5.6

Table A.11. External Public Debt and Debt Service Payments of 84 LDCs (Continued)

Country	Total Debt	Of Which: Official	Of Which: Banks	Of Which: Other Private	Debt Actually Disbursed	Debt Service as Percent of Exports[a] 1973	1974	1975
Sierra Leone	198.3	118.6	—	79.7	163.3	10.0	8.9	9.9
Somalia	424.2	420.2	—	3.9	265.9	3.7	5.0	4.0
Sudan	1,534.8	948.8	406.4	179.7	941.7	12.2	14.8	21.3
Swaziland	66.5	65.0	—	1.5	46.2	9.4	2.4	1.1
Tanzania	1,192.5	1,138.0	12.3	42.7	789.9	7.7[b]	5.7[b]	7.3[b]
Togo	156.7	83.0	44.5	29.2	103.4	7.1	3.8	8.5
Uganda	218.4	210.0	2.2	6.2	185.0	6.8[b]	5.7[b]	10.5[b]
Upper Volta	186.0	185.4	—	0.6	62.9	5.6	4.7	5.2
Zaire	2,738.5	1,043.8	1,273.7	421.0	1,684.2	8.5	11.7	16.5
Zambia	1,464.7	894.2	407.1	163.4	953.4	13.8[c]	5.2	8.1
East Asia and Pacific								
Fiji	80.4	65.9	—	14.4	49.2	0.9	1.7	1.6
Korea	6,996.3	3,678.0	1,376.5	1,941.9	5,227.1	12.2	11.6	11.6
Malaysia	3,020.6	1,355.4	1,561.6	103.7	1,270.1	2.0	2.3	3.2
Philippines	2,567.4	1,807.6	572.5	187.4	1,282.6	8.8	5.1	7.6
Singapore	781.7	481.3	30.8	269.5	521.6	0.6	0.5	0.6
Taiwan	3,137.3	1,630.0	729.9	777.3	1,694.7	3.6	2.7	4.0
Thailand	1,293.2	1,225.4	45.8	22.0	615.5	2.7	2.0	2.6
Latin America								
Argentina	4,698.5	1,816.9	872.2	2,009.4	2,970.5	17.6	16.1	22.0
Bolivia	1,202.7	804.7	164.2	233.8	779.9	15.5	11.5	15.2
Brazil	14,121.0	5,932.9	6,424.3	1,763.7	11,459.3	13.2	13.0	15.0
Chile	4,142.7	2,486.9	422.1	1,233.7	3,559.6	10.9	11.7	28.4

Table A.11. External Public Debt and Debt Service Payments of 84 LDCs (Continued)

Country	Total Debt	Of Which: Official	Of Which: Banks	Of Which: Other Private	Debt Actually Disbursed	Debt Service as Percent of Exports[a]: 1973	1974	1975
Colombia	2,975.0	2,228.1	415.7	431.3	2,355.9	13.6	17.0	11.8
Costa Rica	691.1	490.0	157.4	43.7	410.2	10.3	9.7	10.7
Dominican Rep.	710.1	506.6	166.2	37.3	439.7	5.6	6.0	6.1
El Salvador	395.8	303.2	87.1	5.5	204.9	5.3	4.8	3.4[c]
Guatemala	288.1	273.6	8.3	6.3	172.1	3.6	3.9	1.7
Guyana	361.7	245.7	50.9	65.1	231.6	5.7	4.9	4.6
Honduras	440.1	416.6	11.9	11.6	272.7	3.8	3.7	4.6
Jamaica	812.7	323.1	413.9	75.6	647.4	5.4	6.6	7.3
Mexico	13,480.1	3,649.8	7,560.9	2,269.4	11,251.0	24.3	19.2	25.9
Nicaragua	782.4	422.9	234.6	124.9	600.6	17.4	10.2	11.3
Panama	961.0	372.1	519.0	70.0	655.7	15.1	8.4	8.8
Paraguay	368.8	288.0	43.1	37.6	175.6	8.6	6.7	7.9
Peru	3,466.1	1,378.4	1,553.9	433.8	2,670.1	23.6[c]	18.1[c]	23.8
Uruguay	1,022.9	487.3	218.9	316.8	614.5	22.9	32.8	45.9
Middle East and North Africa (non-OPEC)								
Egypt	5,686.9	5,236.2	268.6	182.1	3,639.7	34.1	30.5	25.7
Jordan	564.8	542.5	8.6	13.6	332.6	8.0	6.3	5.1
Morocco	2,461.9	1,619.4	539.5	302.9	1,593.4	10.1	6.5	7.0
Syria	1,845.5	3,241.5	30.2	193.3	660.7	7.5	6.6	8.2
Tunisia	1,737.6	1,503.9	126.9	106.9	1,071.6	12.0	7.2	7.6
South Asia								
Afghanistan	1,564.0	1,561.9	—	2.1	786.7	19.1	17.3	9.5
Bangladesh	2,582.5	2,403.9	15.3	163.3	1,621.7	2.2	6.0	17.9

Table A.11. External Public Debt and Debt Service Payments of 84 LDCs (Continued)

Country	Total Debt	Of Which: Official	Of Which: Banks	Of Which: Other Private	Debt Actually Disbursed	Debt Service as Percent of Exports[a]: 1973	1974	1975
Burma	531.1	494.2	—	36.8	278.8	26.7	13.3	18.0
India	14,580.1	14,126.7	98.6	354.8	11,538.7	19.9	17.5	16.3
Pakistan	6,328.1	6,081.3	100.2	146.5	5,095.7	16.9	15.8	19.1
Sri Lanka	997.4	863.1	—	134.5	600.7	12.9	12.1	20.3
More Advanced Mediterranean Countries								
Cyprus	99.1	77.7	5.5	15.9	76.3	2.1	2.5	3.5
Greece	2,973.1	766.8	1,885.6	320.7	2,540.5	10.1	9.7	11.6
Malta	48.2	45.9	2.2	—	31.7	0.5	0.8	0.7
Portugal	966.5	250.7	329.9	385.8	665.9	3.3	2.6	4.4
Spain	3,922.9	1,374.6	1,976.6	580.7	3,333.1	3.7	2.6	2.9
Turkey	4,565.0	4,155.8	233.8	165.5	3,180.8	13.4	11.0	11.5
Yugoslavia	3,353.1	2,896.7	220.8	235.6	2,272.5	6.8	6.7	6.7
Oil Exporters (OPEC)								
Algeria	9,003.1	2,015.7	3,477.3	3,510.1	4,526.9	15.4	5.5[c]	9.3
Ecuador	753.4	510.3	136.9	106.0	429.5	7.4	7.6	4.6
Gabon	515.1	146.3	241.9	127.0	454.0	14.7	4.8	6.5
Indonesia	11,764.3	7,123.0	2,516.6	2,124.7	8,124.1	6.6	3.8	7.1
Iran	7,126.3	4,368.7	1,047.4	1,710.2	4,617.3	13.3	5.2	4.4
Iraq	1,132.0	619.1	500.0	12.9	596.1	3.0	1.6	1.0

Table A.11. External Public Debt and Debt Service Payments of 84 LDCs (Continued)

Country	Total Debt	Of Which: Official	Banks	Other Private	Debt Actually Disbursed	Debt Service as Percent of Exports[a]: 1973	1974	1975
Nigeria	1,248.8	1,218.9	12.9	17.0	813.0	4.1	1.7	2.9
Trinidad-Tobago[d]	212.9	135.9	62.2	14.8	155.5	3.0	2.6[c]	1.3[c]
Venezuela	1,379.4	645.0	289.7	444.7	1,222.4	6.3	3.6[c]	4.3[c]
Total	$173,912.1	$107,310.0	$40,557.7	$26,043.5	$120,982.4			

Notes: Components may not add because of rounding.

[a] Exports of goods and non-factor services.

[b] Includes one-third of East African Community debt service.

[c] Adjusted for debt prepayments.

[d] Non-OPEC oil exporter.

Source: *IMF Survey* (5 September 1977).

Table A.12. LDC Cereal Production and Trade, 1948–1952 to 1970–1972 (1,000's tons)

	1948–1952		1959–1961		1970–1972	
	Production	Net Movement	Production	Net Movement	Production	Net Movement
All Cereals						
Africa	26,380	– 300	46,620	– 2,970	43,579	– 4,120
Asia[a]	150,320	– 5,900	176,913	–11,630	276,800	–36,530
Latin America	30,930	+ 2,100	36,887	+ 2,000	70,264	+ 1,440
Total LDCs[b]	207,630	– 4,100	260,420	–12,600	390,643	–39,210
Net Movement/Production Ratio		–2.0%		–4.8%		–10.0%
Barley						
Africa	3,190	– 543	2,523	– 71	4,637	0
Asia[a]	9,790	– 238	11,903	– 625	11,915	– 1,849
Latin America	1,300	– 259	1,096	+ 184	1,484	– 26
Total LDCs[b]	14,280	– 1,040	16,122	– 512	18,036	– 1,875
Net Movement/Production Ratio		–7.3%		–3.2%		–10.4%
Maize						
Africa	8,560	– 51	12,183	+ 873	11,334	– 95
Asia[a]	7,050	+ 232	11,900	– 1,158	17,874	– 6,213
Latin America	14,950	+ 1,140	23,173	+ 2,345	37,054	+ 4,866
Total LDCs[b]	30,560	+ 1,321	47,256	+ 2,060	66,262	– 1,442
Net Movement/Production Ratio		+4.3%		+4.4%		–2.2%

Table A.12. LDC Cereal Production and Trade, 1948–1952 to 1970–1972 (Continued)

	1948–1952		1959–1961		1970–1972	
	Production	Net Movement	Production	Net Movement	Production	Net Movement
Rice						
Africa	3,360	– 83	4,350	– 294	4,896	– 733
Asia[a]	97,070	– 51	135,327	+ 864	171,903	– 517
Latin America	4,720	– 112	7,953	– 145	11,548	– 90
Total LDCs[b]	105,150	– 246	147,630	– 425	188,347	– 1,340
Net Movement/Production Ratio		–0.2%		–0.3%		–0.7%
Wheat						
Africa	4,290	– 925	4,976	– 1,891	5,482	– 2,418
Asia[a]	21,280	– 3,810	30,817	–10,026	53,389	–19,636
Latin America	7,960	+ 150	8,720	– 1,256	11,804	– 3,653
Total LDCs[b]	33,530	– 4,585	44,513	–13,173	70,675	–25,707
Net Movement/Production Ratio		–13.6%		–29.6%		–36.4%
Wheat Flour						
Africa	N.A.	– 943	N.A.	– 827	N.A.	– 541
Asia[a]	N.A.	– 76	N.A.	– 1,376	N.A.	– 1,785
Latin America	N.A.	– 714	N.A.	– 631	N.A.	– 613
Total LDCs[b]	N.A.	– 1,733	N.A.	– 2,834	N.A.	– 2,939

Table A.12. LDC Cereal Production and Trade, 1948-1952 to 1970-1972 (Continued)

	1948-1952		1959-1961		1970-1972	
	Production	Net Movement	Production	Net Movement	Production	Net Movement

Notes: Production and Net Movements are annual averages for indicated periods.

N.A. indicates not available.

[a] Omitting People's Republic of China.

[b] LDC category slightly different than elsewhere in this paper.

Source: *FAO Production Yearbook* and *FAO Trade Yearbook* (various issues).

APPENDIX II
RECENT DEVELOPMENTS IN OPEC ECONOMIC ASSISTANCE PROGRAMS

This section comprises a partial listing of recent OPEC economic aid to the LDCs. The programs represent a fairly complete compilation of the activities of these specific institutions for the indicated periods, but comparably recent data are not yet available for other major OPEC donors, such as Iraq, Libya, and Venezuela. Nevertheless, these figures represent the scope and direction of OPEC aid in the period following that covered by the more complete OECD data quoted earlier (see above, chapter 3).

(1) ABU DHABI FUND FOR ARAB ECONOMIC DEVELOPMENT (ADFAED)

The ADFAED was founded in July 1971 and was capitalized initially for approximately $125 million. In 1974, when the first loan was actually made, the law governing the Fund was amended to allow loans to non-Arab nations in Africa and Asia. At the same time, capitalization was raised to about $500 million. Through the end of 1975, the ADFAED had committed over $100 million to 13 projects in 6 Arab countries. In 1976, ADFAED activity was up sharply to about $170 million in loan commitments, about a third of which went to 7 non-Arab applications, as is shown below.

Table A.13. ADFAED Economic Assistance, 1976

Beneficiary		Purpose	Amount ($ million)	Amortization Period (years)	Interest Rate (percent)	Approval Date
Bahrain	(1)	Desalination Project	25.0	19	3.5	Feb. 1976
	(2)	Industrial Development	15.0	14	4.5	Feb. 1976
Egypt		Fertilizer Plant	4.6	16	4.0	June 1976
Morocco	(1)	Office Building	4.0	17	3.0	Jan. 1976
	(2)	Textile Mill	17.5	11	5.5	April 1976
Oman		Gas Utilization Project	15.0	20	4.0	July 1976
Sudan	(1)	Textile Mill	20.0	12	5.0	July 1976
	(2)	Rural Development	4.1	20	3.5	Dec. 1976
Y.P.D.R.	(1)	Fisheries	7.25	13	3.5	April 1976
	(2)	Geological Survey	1.6	13	2.5	April 1976

Table A.13. ADFAED Economic Assistance, 1976 (Continued)

Beneficiary	Purpose	Amount ($ million)	Amortization Period (years)	Interest Rate (percent)	Approval Date
Bangladesh	Machinery Plant	10.0	13	4.5	Nov. 1976
Burundi	Fisheries	1.0	21	2.5	June 1976
India	Hydroelectric Project	17.0	19	3.5	July 1976
Malaysia	(1) Flour Mill	4.0	14	4.0	Nov. 1976
	(2) Palm Oil Project	4.25	14	4.0	Nov. 1976
Maldive Is.	Fisheries	2.0	12	3.0	Aug. 1976
Mali	Selingue Dam	4.3	20	3.0	Oct. 1976
Sri Lanka	Fisheries	5.0	13	5.0	Nov. 1976

All loan repayments involve a grace period of three to six years. In addition to these projects, by early 1978 the ADFAED was unofficially reported to be negotiating another 33 loans for approximately $310 million. As can be seen below, these proposals involve 14 more non-Arab nations not previously recipients of ADFAED loans, with the latter reported to be receiving consideration for about 60 percent of the pending ADFAED funding commitments.

Table A.14. ADFAED Pending Projects, 1977–1978

Applicant		Purpose	Amount ($ millions)
Bahrain	(1)	Industrial Development	15.3
	(2)	Poultry Project	5.0
	(3)	Electric Power	5.0
Egypt	(1)	Suez Canal	15.4
	(2)	Tourist Hotel	4.5
Jordan		Phosphate Project	25.6
Mauritania	(1)	Textile Factory	14.3
	(2)	Iron Mine	
Sudan		Railway	10.1
Syria		Electric Power	14.3
Tunisia		Transport Project	12.0
Y.A.R.		Topographic Study	1.3
Y.P.D.R.	(1)	Geological Survey	1.8
	(2)	Electric Power	20.5
Afghanistan		Sugar Plant	7.6
Bangladesh		Electric Power	12.0
Cape Verde		Fisheries	1.0
Comoro Is.		Electric Power	1.0
Gambia		Airport	1.3
Guinea		Cement Plant	4.1
Indonesia		Electric Power	14.6
Lesotho		Airport	0.8
Malta		Port Development	7.0
Mozambique	(1)	Irrigation Project	40.0
	(2)	Highways	
Niger		Livestock Project	2.5
Rwanda		Factory	0.5
Senegal		Barrage Project	5.1
Tanzania	(1)	Fruit Production	7.1
	(2)	Sugar Plant	6.1
	(3)	Sugar Cultivation	30.7
Uganda	(1)	Highways	10.0
	(2)	Textile Plant	6.4

Sources: *Annual Report 1976,* ADFAED; MEED Special Report: United Arab Emirates, *Middle East Economic Digest* (July 1977); *Middle East Economic Digest* (19 August, 21 October, 18 November, and 25 November 1977); *Middle East Economic Survey* (22 August, 14 November, 8 December 1977, and 2 January, 9 January, 22 February, and 6 March, 1978).

(2) ARAB BANK FOR ECONOMIC DEVELOPMENT IN AFRICA (ABEDA)

In 1976, the Special Arab Aid Fund for Africa (SAAFA), set up in the first flash of oil-exporter concern for their erstwhile compatriots in poverty, was merged with ABEDA. Following this, the Bank's total capitalization amounted to $631 million, with the subscriptions as listed below.

Table A.15. ABEDA Capitalization

Country	Amount ($ millions)	Voting Power (percent)
Algeria	20	4.04
Bahrain	1	2.12
Egypt	1	2.12
Iraq	100	12.11
Jordan	1	2.12
Kuwait	100	12.11
Lebanon	5	2.52
Libya	100	12.11
Mauritania	1	2.12
Morocco	10	3.03
Oman	14	3.43
Palestine	1	2.12
Qatar	40	6.05
Saudi Arabia	150	17.15
Sudan	1	2.12
Syria	1	2.12
Tunisia	5	2.52
U.A.E.	80	10.09

In February 1977, the Board of Governors voted to increase ABEDA's capital by another $115 million.

Before SAAFA was transferred to ABEDA management, it had granted $152.5 million in emergency aid to 32 countries. Amounts ranged from $250 thousand (Guinea-Bissau) and $500 thousand (each to Cape Verde, Comoro Islands, Equatorial Guinea, and Sao Tome-Principe) to $14.2 million (each to Ethiopia and Tanzania). All loans are to be repaid in 25 years, including a 10-year grace period, and are subject to a one percent annual administrative charge. SAAFA was initially envisaged as a special purpose fund, analogous to the IMF's Oil Facility, lending to those African states whose international financial situations were most distressed in the months following the increase in oil prices. Its sponsors, acting through the League of Arab States, decided that their immediate response to economic problems of continuing duration were best handled, over the long-run, through ABEDA, which, though founded at the same time as SAAFA, did not begin formal operations until late 1975.

Since then, ABEDA funding has been made available for specific projects and with terms varying from case to case, as can be seen below. By early 1978, 30 sub-Saharan countries had received ABEDA loans.

Table A.16. ABEDA Economic Assistance, 1975–1978

Beneficiary		Purpose	Amount ($ millions)	Amortization Period (years)	Interest Rate (percent)	Approval Date
Angola		SAAFA Emergency Aid	13.2	25	1	Mar. 1977
Benin		Cement Plant	8.0	25	6	Nov. 1975
Burundi	(1)	Sewerage System	4.0	25	2	June 1976
	(2)	Highway	6.0	20	2	Mar. 1978
Cameroon	(1)	Port Facility	10.0	25	4	Nov. 1975
	(2)	Paper Pulp Plant	10.0	11	7	Nov. 1977
Cape Verde		SAAFA Emergency Aid	10.0	25	1	Oct. 1976
Comoro Is.		SAAFA Emergency Aid	10.0	25	1	Aug. 1976
Congo		Railway	10.0	25	4	Nov. 1975
Gambia		Rural Development	3.3	25	2	Mar. 1976
Ghana	(1)	Cocoa Project	5.0	25	4	Nov. 1975
	(2)	Kpong Dam	10.0	17	6	Mar. 1977
Guinea		Cement Plant	4.8	17	5	July 1977
Kenya		Rural Development	5.0	25	4	Mar. 1976
Lesotho		Airport	6.0	18	4	Mar. 1978
Liberia	(1)	Industrial Estate	3.2	14	5	Nov. 1975
	(2)	Electric Power	3.9	13	7	Mar. 1978
Madagascar	(1)	Highway	5.0	25	3	Nov. 1975
	(2)	Electric Power	10.0	20	4	Nov. 1977
Mali	(1)	Selingue Dam	15.0	25	2	Mar. 1976
	(2)	Rural Development	5.0	25	2	Mar. 1977
Mauritius		Electric Power	10.0	15	4	June 1976
Mozambique	(1)	SAAFA Emergency Aid	6.0	25	1	July 1976
	(2)	SAAFA Emergency Aid	20.0	25	1	Sept. 1976

Table A.16. ABEDA Economic Assistance, 1975-1978 (Continued)

Beneficiary	Purpose	Amount ($ millions)	Amortization Period (years)	Interest Rate (percent)	Approval Date
Niger	Highway	7.0	25	2	Nov. 1976
Rwanda	(1) Rural Development	5.0	25	2	June 1976
	(2) Electric Power	6.0	20	4	July 1977
Sao Tome & Principe	SAAFA Emergency Aid	10.0	25	1	Sept. 1976
Senegal	(1) Livestock Project	1.6	25	2	Nov. 1975
	(2) Fishing Port	7.2	20	5	Mar. 1977
Sierra Leone	Electric Power	5.0	15	4	Mar. 1976
Tanzania	(1) Maize Project	5.0	20	2	Nov. 1975
	(2) Construction	10.0	15	7	July 1977
Uganda	Textile Plant	4.7	15	4	Mar. 1978
Upper Volta	(1) Railway	10.0	25	2	Nov. 1975
	(2) Rural Development	4.5	25	2	June 1976
Zambia	Highway	10.0	25	4	June 1976
Zaire	Water Supply	10.0	25	4	Nov. 1975
Multinational (Ghana, Ivory Coast, Togo)	Cement Plant	10.0	15	6	Nov. 1975

Note: N.A. indicates data not available.

Sources: *Annual Report 1976,* ABEDA; *BADEA et FASA Prets a jusqu'an 30 Juin 1977,* ABEDA; *Middle East Economic Survey* (25 July, 1 August, and 12 December, 1977).

(3) ARAB FUND FOR ECONOMIC AND SOCIAL DEVELOPMENT (AFESD)

The AFESD was founded in 1968, but did not actually begin operations until 1973, and then only with the guidance of the Kuwait Fund for Arab Economic Development. During 1976, its subscribed capital was raised from $355 million to $660 million. Eventually all members of the Arab League became subscribers to the AFESD, with voting shares distributed, as indicated below, in proportion to their contributions (10 thousand Kuwaits dinars per share) to the Fund's capital.

Table A.17. AFESD Shareholders

Country	Shares
Algeria	400
Bahrain	50
Egypt	1,050
Iraq	750
Jordan	200
Kuwait	3,000
Lebanon	100
Libya	1,200
Mauritania	10
Morocco	200
Oman	204
Palestine	5
Qatar	100
Saudi Arabia	1,880
Somalia	5
Sudan	150
Syria	300
Tunisia	50
U.A.E.	500
Y.A.R.	50
Y.P.D.R.	1
Total	10,205

Through the end of 1975, the AFESD had committed about $325 million; but during 1976 alone, more than $340 million in additional loans were negotiated. Unlike other OPEC-related aid institutions, the AFESD has continued to cater only to the development needs of Arab countries, as can be seen below.

Table A.18. AFESD Economic Assistance, 1976

Beneficiary		Purpose	Amount ($ millions)	Amortization Period (years)	Interest Rate (percent)	Approval Date
Egypt	(1)	Fertilizer Project	9.4	18	4	June 1976
	(2)	Electric Power	41.8	20	4	Nov. 1976
	(3)	Textile Plant	34.8	20	4	Dec. 1976
Jordan		Electric Power	20.9	20	6	April 1976
Mauritania	(1)	Electric Power	18.1	25	4	April 1976
	(2)	Highway	24.4	25	4	Sept. 1976
Morocco		Irrigation	24.4	20	6	April 1976
Oman		Gas Utilization	20.9	15	6	Nov. 1976
Somalia		Irrigation	22.3	25	4	June 1976
Sudan		Highway	38.3	20	4	Dec. 1976
Syria		Drinking Water	41.8	15	6	Sept. 1976
Y.A.R.		Highway	13.2	25	4	June 1976
Y.P.D.R.	(1)	Highway	22.6	25	4	June 1976
	(2)	Multipurpose Project	9.1	18	4	Sept. 1976

Note: The grace period for most loans was five years.

AFESD data relative to 1977 is tentative; 16 transactions with 11 countries and 2 international institutions have been indicated. Recipients included: Algeria ($42 million for port construction); Bahrain ($17 million for electric power); Egypt ($42 million for Suez Canal Development); Jordan ($21.5 million for electric power); Lebanon ($21 million for repairs to the electric power grid and $17 million for Beirut port facilities); Morocco ($31 million for agricultural mechanization and $32 million for a cement plant); Somalia ($28 million for highways); Sudan ($17 million for railways); Syria ($21 million for electric power); Tunisia ($23 million for development of several small industrial plants); and Yemen Arab Republic ($17 million for drinking water development and $31 million for electric power). In addition the AFESD pledged $20 million to the Arab Air Transport Union and the Arab Civil Aviation Council for installation of a computer system. Though loan terms were not yet available in each case, this $380 million was believed to have been disbursed under policies quite similar to those followed in 1976.

Sources: *Annual Report 1976*, AFESD; *Middle East Economic Digest* (7 April 1977); *Middle East Economic Survey* (21 March, 4 April, 18 April, 25 April, 13 June, 14 November, 5 December 1977 and 13 February 1978).

(4) ISLAMIC DEVELOPMENT BANK (IDB)

The IDB was founded in December 1973 in response to the serious concern among the more orthodox Muslim oil-exporting countries over the Quar'anic injunctions against the taking of interest. Though the scriptural prohibitions are, in context, absolute, even modernistic interpretations leave considerable difficulty with regard to two specific categories: loans to Muslims and to the poor, of whatever faith. In our context here, then, the concern is with loans to the non-oil LDCs with sizable, though not necessarily majority, Muslim populations. The Qua-r'an advises *gifts* to the poor, but does not rule out loans, with repayment only of the principal; the IDB founding members interpreted this to allow minimal service charges, attached to loans in order to defray administrative costs.

By mid-1978, IDB had 34 subscribing members and an authorized capital of 2,000 million Islamic dinars ($2,325 million; 1 ID equals 1 SDR, or in early 1978 about $1.16); paid-up capital equaled about $875 million in convertible currency. Loans could be made only to members (Table A.19). The first loan was extended in October 1976, and by the end of the 1976-1977 fiscal year (in December 1977), some $182.3 million in credits had been extended to 16 members (Table A.20) to help finance 31 development projects. The IDB disbursements, however, lagged far behind commitments during its first year of actual operations; only about $15 million, or 8 percent of total commitments, was actually disbursed by December 1977. In early 1978, another 6 projects were extended assistance of about $40 million, bringing to 19 the number of members who had tapped the IDB fund.

Table A.19. IDB Membership Subscriptions

Member	Amount (ID million)	Member	Amount (ID million)
Afghanistan	2.5	Niger	2.5
Algeria	25.0	Oman	5.0
Bahrain	5.0	Pakistan	25.0
Bangladesh	10.0	Palestine	2.5
Cameroon	2.5	Qatar	25.0
Chad	2.5	Saudi Arabia	200.0
Egypt	25.0	Senegal	2.5
Guinea	2.5	Somalia	2.5
Indonesia	25.0	Sudan	10.0
Jordan	4.0	Syria	2.5
Kuwait	100.0	Tunisia	2.5
Lebanon	2.5	Turkey	10.0
Libya	125.0	Uganda	2.5
Malaysia	16.0	U.A.E.	110.0
Mali	2.5	Upper Volta	2.5
Mauritania	2.5	Y.A.R.	2.5
Morocco	5.0	Y.P.D.R.	2.5
		Total	767.5

Table A.20. IDB Economic Assistance, 1976–1978

Beneficiary		Purpose	Amount ($ millions)	Amortization Period (years)	Approval Date
Algeria	(1)	Trade Financing	12.7	N.A.	Feb. 1977
	(2)	Port Development	7.5	20	Dec. 1977
Egypt	(1)	Suez Canal	12.0	18	Sept. 1977
	(2)	Sugar Project	6.0	N.A.	Jan. 1978
Jordan		Oil Refinery[a]	9.3	N.A.	Dec. 1976
Mauritania		University	0.8	30	Dec. 1977
Morocco		Cement Plant[a]	10.0	N.A.	Sept. 1977
Somalia		Highway	7.5	30	Mar. 1977
Sudan	(1)	Textile Mill[a]	2.5	N.A.	Feb. 1977
	(2)	Highway	5.0	30	May 1977
	(3)	Educational Equipment	3.5	21	May 1977
	(4)	Trade Financing	3.2	N.A.	Sept. 1977
Syria		Electric Power	6.5	N.A.	Jan. 1978
Tunisia		Fisheries	7.0	30	June 1977
U.A.E. (Dubai)		Housing	5.0	N.A.	Feb. 1978
Y.P.D.R.		University	2.0	30	Oct. 1977
Afghanistan		Satellite Communications	8.0	N.A.	Mar. 1978
Bangladesh	(1)	Airport	7.0	30	Mar. 1977
	(2)	Shipping	6.0	N.A.	Dec. 1977
Cameroon	(1)	Electric Power	7.0	25	Oct. 1976
	(2)	Paper Pulp Project[a]	8.3	N.A.	Sept. 1977
Guinea		Cement Plant[a]	4.7	N.A.	Aug. 1977
Malaysia	(1)	Cement Plant[a]	6.7	N.A.	Aug. 1977
	(2)	Development Bank[a]	5.75	N.A.	Sept. 1977
	(3)	Development Studies	1.8	N.A.	Feb. 1978
Pakistan	(1)	Oil Refinery[a]	4.0	N.A.	June 1977
	(2)	Trade Financing	15.0	N.A.	Aug. 1977
	(3)	Textile Plant[a]	2.3	N.A.	Sept. 1977
Senegal		Highway	6.25	25	Dec. 1977
Turkey	(1)	Trade Financing	10.0	N.A.	June 1977
	(2)	Trade Financing	9.6	N.A.	Dec. 1977
	(3)	Tractor Plant	12.0	N.A.	Feb. 1978

Note: N.A. indicates data not available or not applicable.

[a] Equity Participation

Sources: *Middle East Economic Survey* (30 January, 16 February, 17 March, and 24 April 1978).

(5) KUWAIT FUND FOR ECONOMIC DEVELOPMENT (KFAED)

The KFAED was the first major venture by an oil exporter into the field of granting and overseeing foreign assistance. Founded in December 1961, the KFAED, as its name implies, was intended to channel aid to Kuwait's Arab neighbors. Through the beginning of the 1975-1976 fiscal year, over $560 million was actually disbursed to 14 Arab countries, with Egypt ($120 million), the Sudan ($77 million), Tunisia ($67 million), and Syria ($66 million) the major beneficiaries during this period. Beginning in April 1975, the KFAED charter was amended to allow for lending to non-Arab countries, and during the following 15-month fiscal year, loans totalling more than $290 million were disbursed to Arab borrowers, while non-Arab states were lent about $261 million, as can be seen below.

By the end of the 1976-1977 fiscal year (30 June 1977), it has been estimated that the cumulative Kuwaiti financial aid (including, but not limited to, KFAED activities) to LDCs amounted to nearly $5,000 million, though not all of this had actually been distributed.[1] Of this, about 55 percent had been distributed as grants and 45 percent as loans.

Table A.21. KFAED Economic Assistance, 1975–1976

Beneficiary		Purpose	Amount ($ millions)	Amortization Period (years)	Interest Rate (percent)	Approval Date
Egypt	(1)	Gas Project	12.2	20	4	July 1975
	(2)	Electric Power	34.8	20	4	Oct. 1975
	(3)	Fertilizer Project	11.1	18	4	June 1976
Jordan		Industrial Development Bank	8.7	15	4	May 1976
Mauritania	(1)	Highway	4.0	25	0.5	April 1975
	(2)	Highway	20.2	25	0.5	Oct. 1975
	(3)	Port Expansion	8.5	25	1.5	Feb. 1976
Morocco	(1)	Industrial Development Bank	20.9	16	7	Aug. 1975
	(2)	Electric Power	8.7	17	4	Sept. 1975
Somalia	(1)	Electric Power	21.6	29	1.5	July 1975
	(2)	Rangeland Development	20.9	39	1.5	June 1976
Sudan	(1)	Industrial Development Bank	5.2	16	4	June 1975
	(2)	Irrigation	39.0	31	3	June 1975
Tunisia	(1)	Industrial Development Bank	8.7	16	4	April 1975
	(2)	Airport	13.9	20	4	Oct. 1975
	(3)	Highway	13.0	20	4	Feb. 1976
	(4)	Bridge	7.8	19	4	Feb. 1976
Y.A.R.	(1)	Irrigation	9.7	47	0.5	Jan. 1976
	(2)	Highway	5.2	39	1.5	Jan. 1976
Y.P.D.R.		Highway	15.7	40	1.5	June 1975

Table A.21. KFAED Economic Assistance, 1975–1976 (Continued)

Beneficiary	Purpose	Amount ($ millions)	Amortization Period (years)	Interest Rate (percent)	Approval Date
Bangladesh	(1) Irrigation	8.0	33	1.5	June 1975
	(2) Irrigation	22.3	32	2	June 1975
Comoro Is.	Highway	6.3	39	1	June 1976
Guinea	Communications	9.4	18	4	June 1976
India	Electric Power	52.2	25	4	Jan. 1976
Malaysia	Rural Development	26.4	30	5	June 1975
Maldive Is.	Airport	5.2	20	1.5	June 1976
Nepal	Kulekhani Dam	17.4	32	3	Jan. 1976
Pakistan	Electric Power	45.2	24	4	April 1976
Rwanda	Tea Production Project	3.5	27	3	June 1975
Sri Lanka	Fertilizer Project	26.1	19	4	Oct. 1975
Tanzania	Textile Plant	15.7	22	4	July 1975
Thailand	Electric Power	3.5	24	4	April 1976
Uganda	Livestock Project	20.0	25	2.5	June 1975

The expanded financial commitment and geographical scope of the Kuwaiti economic assistance program is further emphasized by KFAED's list of projects under active consideration for implementation after 1 July 1976. Several had already been the subject of KFAED-financed feasibility studies. Included on the list were 15 additional potential non-Arab recipients.

Table A.22. KFAED Pending Projects, 1976–1977

Potential Beneficiary		Purpose
Bahrain		Electric Power
Egypt	(1)	Oil Pipeline
	(2)	Suez Canal Widening
	(3)	Textile Plant
Jordan	(1)	Phosphate Project
	(2)	Livestock Project
	(3)	Thermal Power Project
Lebanon		Airport
Morocco		Al-Makhazen Dam
Oman	(1)	Gas Pipelines
	(2)	Livestock Project
Somalia		Highway
Sudan	(1)	Planning Studies
	(2)	Highway
	(3)	Railway
	(4)	Sugar Project
Syria		Highway
Tunisia	(1)	Railway
	(2)	Waterway
Y.A.R.	(1)	Livestock Project
	(2)	Cement Plant
	(3)	Hodeida Port
	(4)	Water Resources

Table A.22. KFAED Pending Projects (Continued)

Potential Beneficiary		Purpose
Y.P.D.R.	(1)	Port Facilities
	(2)	Airport
	(3)	Fishmeal Plant
	(4)	Grain Silos
	(5)	Hospital
Afghanistan		Sugar Plant
Bangladesh		Water Survey
Burundi	(1)	Coffee Production
	(2)	Highway
Cameroon		Hydroelectric Project
Congo		Railways
Cyprus		Highways
Gambia		Highways
Ghana		Hydroelectric Project
Guinea		Highway
Guinea-Bissau		Port and Airport
Indonesia		Electric Power
Lesotho		General Development
Madagascar	(1)	Highway
	(2)	Electric Power
Malaysia		Palm Oil Project
Malta		Fisheries
Mali		Selingue Dam
Pakistan		Railway
Philippines		Electric Power
Senegal		Livestock Project
Tanzania		Textile Plant Extension

By early 1978, several of these projects were reported, at least unofficially, to have been approved and funded; among them: Bahrain ($22 million); Egypt ($24 million for pipeline construction, $50 million for the Suez Canal, and $34 million for a textile plant); Jordan ($25 million for phosphates and $21 million for thermal power); Lebanon ($28 million); Morocco ($35 million); Oman ($26 million for pipeline and $0.5 million for livestock); Somalia ($19 million); Sudan ($3.1 million for planning, $17 million for railways, $32 million for highways, and $21 million for sugar); Syria ($14 million); Tunisia ($25 million for waterway development); Yemen Arab Republic ($12 million for livestock improvement projects, $8.4 million for Hodeida port, and $1.3 million for water resource development); Yemen People's Democratic Republic ($9 million); Afghanistan ($31 million); Bangladesh ($56 million); Burundi ($1.25 million for coffee and $6.2 million for highways); Cameroon ($16 million); Congo ($14 million); Cyprus ($4 million); Gambia ($16 million); Ghana ($31 million); Indonesia ($30.5 million); Madagascar ($7.3 million for highways and $10.3 million for electric power); Malaysia ($8.5 million); Mali ($17 million); Malta ($4 million); Pakistan ($25 million); the Philippines ($12 million); Senegal ($4 million); and Tanzania ($10 million). Of these nearly $710 million in commitments, nearly half went to non-Arab countries.

Sources: *Annual Report 1975-1976 and Annual Report 1976-1977,* KFAED; MEED Special Report; Kuwait, *Middle East Economic Digest* (August 1977); *Middle East Economic Digest* (21 January, 18 February, and 22 December 1977, and 6 January 1978); *Middle East Economic Survey* (29 November 1976, 10 January, 28 February, 4 April, 25 April, 9 May, 13 June, 27 June, 25 July, 21 November 1977, 27 February and 15 May 1978).

(6) THE OPEC SPECIAL FUND

The OPEC Special Fund was established in January 1976 as a means for coordinating, at least in part, the economic assistance of the member states. To some extent, the institution was founded in response to criticism by the industrialized countries and some LDCs that many oil exporters were rather slow in actually granting aid promised in principle to the LDCs. Since several OPEC members had no aid-giving or aid-administering experience before 1974, it was argued that a multilateral organization offered the possibility of more efficient management of assistance programs. At the same time, since the Special Fund was affiliated with OPEC, it would have the political advantage of clearly identifying to beneficiaries where the assistance originated, whereas aid channelled through existing international organizations would assume an anonymity that might be unattractive to some OPEC donors.

Though initial discussions were in terms of a capitalization of $1,000 million, agreement on specific details was hard to reach. The first achievement of the Special Fund's governing board was in coordinating in May 1976 the response of OPEC to the foundation of the International Fund for Agricultural Development (IFAD) as a specialized agency of the United Nations. The eventual contributions agreed upon by the members are (in millions of dollars) indicated below.

Table A.23. OPEC Special Fund Contributors to IFAD ($ millions)

Algeria	10.00
Gabon	0.50
Indonesia	1.25
Iran	124.75
Iraq	20.00
Kuwait	36.00
Libya	20.00
Nigeria	26.00
Qatar	9.00
Saudi Arabia	105.50
U.A.E.	16.00
Venezuela	66.00

As can be seen, the participation in IFAD of several states with fairly extensive bilateral aid programs of their own, like Kuwait, Libya, and the U.A.E., is relatively less prominent than that of others, like Venezuela and Nigeria, whose bilateral programs have been smaller.

The second source of contributions for the Special Fund came in response to the agreement on a $200 million balance of payments support program for the Most Seriously Affected LDCs (MSAs). These initially came to 45, but their ranks grew to 49 by early 1977. Commitments to MSAs were based on four criteria: per capita income, decline in gross foreign exchange reserves, ratio of imports and debt service payments to reserves, and increase in oil import payments. Action on the 49 loans of the Fund's first year were complete by May 1977.

Table A.24. OPEC Special Fund Economic Assistance, 1976 ($ millions)

Country	Amount	Country	Amount
Sudan	7.45	Honduras	1.75
W. Samoa	1.60	Tanzania	5.45
Sri Lanka	8.10	Uganda	4.55
Guinea	2.35	Egypt	14.45
Pakistan	21.45	Y.A.R.	2.25
Cent. Afric. Emp.	1.75	Upper Volta	2.25
Madagascar	3.10	Kenya	5.00
Nepal	4.15	Rwanda	1.70
Mali	3.55	Senegal	3.40
Guinea-Bissau	1.65	Chad	2.40
Cape Verde	1.55	Ghana	7.80
Gambia	1.65	Laos	2.15
Mauritania	1.60	Ethiopia	4.80
Y.P.D.R.	2.40	Sierra Leone	2.05
Guatemala	1.75	Niger	2.90
Guyana	1.60	Burundi	1.70
Haiti	3.15	Cameroon	4.95
Afghanistan	3.75	Mozambique	6.55
India	21.80	Benin	2.00
Bangladesh	13.90	El Salvador	1.75
Lesotho	1.90	Sao Tome	0.35
Burma	2.25	Maldive Is.	0.50
Somalia	2.05	Equat. Guinea	0.50
Grenada	0.35	Comoros	0.50
Seychelles	0.30		
		Total	196.85

All MSA loans carried identical terms: an annual service charge of 0.5 percent and a maturity period of 25 years including a 5-year grace period. Recipients were required to deposit the local currency equivalent of their grant and then, in consultation with the Special Fund's governing board, determine within a year a local development project to be carried out with these counterpart funds.

Finally, the Special Fund is to receive the profits accruing to 8 of its members (Iran, Iraq, Kuwait, Libya, Qatar, Saudi Arabia, U.A.E., and Venezuela) as a result of the IMF's gold auctions. The Fund in turn will donate these profits in OPEC's name to the IMF Trust Fund.

At OPEC meetings in December 1976 and February 1977, it was determined that the Special Fund would be replenished, with a goal of $800 million being set. The Fund's second operating year was devoted to consideration of specific projects to be funded. In August 1977, loans amounting to about $54 million were made to 13 countries, while a donation of nearly $13 million was given to the United Nations Development Program (UNDP) for 4 projects—one each in Latin America, Africa, the Middle East, and East Asia. Another 9 loans to 8 countries were made in November 1977; these totalled $36.5 million, and another $12.7 million was lent to the Asian Development Bank (ADB) for projects in Burma, the Philippines, and Sri Lanka.

(7) SAUDI ARABIAN FUND FOR DEVELOPMENT (SAFD)

Though Saudi Arabia is now far and away the richest and most prominent oil-producing country, it is relatively new to the task of institutionalizing its foreign economic assistance. Until 1974, all such aid was given on an *ad hoc* basis. In August of that year, the SAFD was founded. From the beginning, its charter specified that

Table A.25. OPEC Special Fund Economic Assistance, 1977–1978

Beneficiary		Purpose	Amount ($ millions)	Amortization Period (years)	Interest Rate (percent)
Bangladesh		Airport	3.5	20	0.75
Benin		Highway	1.6	20	0.75
Botswana		Highway	1.0	20	0.75
Burma	(1)	Communications	3.1	19	0.75
	(2)	Water Supply	1.3	20	0.75
Costa Rica		Highway	3.0	19	0.75
Dominican Rep.	(1)	Bao Dam	1.0	20	0.75
	(2)	Fisheries	1.9	20	0.75
Egypt		Development Bank	8.7	20	0.75
Honduras		Electric Power	1.7	19	0.75
India		Oil Fields	14.0	20	0.75
Jamaica		Irrigation	3.0	20	0.75
Jordan		Irrigation	1.7	20	0.75
Kenya		Water Supply	3.0	20	0.75
Malawi		Highway	1.8	20	0.75
Mauritania[a]		Iron Ore Project	5.0	N.A.	0.75
Morocco		Hydroelectric Project	3.0	19	0.75
Nepal[a]		Forestry Development	3.0	20	0.75
Pakistan		Fertilizer Plant	11.0	20	0.75
Philippines		Highway	8.3	20	0.75
Rwanda		Hydroelectric Project	2.4	20	0.75
Sri Lanka		Hydroelectric Project	3.2	20	0.75
Sudan		Highway	3.3	20	0.75
Thailand		Electric Power	7.0	20	0.75
Upper Volta		Highway	2.1	20	0.75
Vietnam[a]		Fisheries	7.0	20	0.75
Zaire[a]		General Development	5.0	N.A.	0.75
ADB		Development Loans	12.7	20	0.75
UNDP	(1)	Energy Program	1.5	grant	0
	(2)	Niger River Dam	5.0	grant	0
	(3)	Red Sea Fisheries	4.2	grant	0
	(4)	Offshore Mineral Exploration	2.0	grant	0

Note: N.A. indicates data not available.

[a] Loan announced in March 1978.

Sources: *First Annual Report 1976* and *Second Annual Report 1977*, The OPEC Special Fund; *Weekly Bulletin Review of the Press*, OPEC; *OPEC Special Fund Press Releases.*

non-Arab, as well as Arab, LDCs were to be eligible for grants. The initial authorized capital was 10,000 million riyals (in early 1978, about $2,890 million, with 3,000 million authorized for commitment in the SADF's first fiscal year (A.H. 1394/95, ending in July 1975), and another 2,300 million allowed for fiscal 1395/96 (ending in June 1976, the lunar basis of the Hijral calendar resulting in a year about 11 days shorter than the solar year.)

However, though the Saudi government intended the SAFD to handle most of its economic assistance, circumstances have slowed the achievement of this goal. On the one hand, already existing Saudi commitments (for example, to international agencies and to the Arab confrontation states) have continued to be the responsibility of other government ministries. On the other hand, Saudi Arabia, like other OPEC donors, has had problems staffing an institution which must not only evaluate aid requests, but also oversee the projects to be financed. With restrictions drawn so as to ensure that SAFD efforts would be widely spread (for example, no more than 5 percent of SAFD capital to be committed to a single project or 10 percent to any one country), the first fiscal year saw the approval of 14 loans to 5 countries totalling about 1.4 billion riyals (about $400 million), slightly less than half the amount originally envisioned for this period. If the announced goals for the SAFD were not completely met in its first year, nevertheless its inauguration marked a notable shift in Saudi Arabia's aid activities. One recent analysis pointed out that, whereas about 60 percent of Saudi aid in 1974 was for general support assistance and 20 percent for project assistance, in 1975 only about 30 percent was for general assistance and 65 percent was tied to specific projects.[2] However, most Saudi aid activity continues to be outside SAFD channels. For example, the Saudi share of the IMF's Witteveen facility came to more than $2,500 million; more than $100 million went to the International Fund for Agricultural Development (IFAD).

Statistics regarding recent SAFD activity are somewhat less complete than those indicated above for Abu Dhabi and Kuwaiti efforts, and the available figures occasionally do not clearly indicate whether a given Saudi loan is entirely within the province of the SAFD. Nevertheless, it is clear that the pace of SAFD activity had stepped up considerably by 1977, as can be seen below in Table A.26. Most loans involve nominal service charges with relatively long payback periods (20 years in many cases), frequently including initial grace periods of 5 years.

Despite the incompleteness of this data, it is probably fair to conclude that most recent SAFD activity is still directed toward Arab League members. Only about $465 million of the some $1,950 shown above went to non-Arab states. Even subtracting the large aid packages for Syria and the Y.A.R. (obviously not to be disbursed immediately), it would seem that about three-fifths of SAFD aid in 1977 and early 1978 went to Arab applicants, a higher proportion than that indicated above for other Arab aid agencies.

Table A.26. SAFD Projects, 1977-1978

Beneficiary		Purpose	Amount ($ millions)	Reported Approval Date
Bahrain		Electric Power Project	43.0	Dec. 1977
Egypt		Suez Canal	49.9	June 1977
Jordan		Aqaba Port	23.1	Oct. 1977
Mauritania		Iron Mining Development	49.0	Dec. 1977
Morocco	(1)	al-Massira Dam	33.9	Jan. 1978
	(2)	Port Expansion	33.9	July 1977
	(3)	Irrigation	32.6	July 1977
Oman	(1)	Highway	74.7	July 1977
	(2)	Regional Development	21.6	July 1977
	(3)	Copper Mining Development	100.0	July 1977
Somalia		Sugar Production Project	69.6	Nov. 1977
Sudan	(1)	Agricultural Development	8.5	July 1977
	(2)	Satellite Communications	10.0	July 1977
	(3)	Airport	7.0	July 1977
	(4)	Educational Equipment	3.9	Oct. 1977
Syria	(1)	Support for Development Plan[a]	156.2	Aug. 1977
	(2)	Electric Power Project[a]	21.2	Dec. 1977
	(3)	University[a]	35.4	Feb. 1978
	(4)	Highway[a]	9.9	Feb. 1978
Tunisia	(1)	Engineering College	10.6	Nov. 1977
	(2)	Drinking Water Project	45.7	Nov. 1977
	(3)	Sidi Sa'd Dam	70.2	Mar. 1978
Y.A.R.	(1)	Electric Power Project[a]	25.0	June 1977
	(2)	Drinking Water Project[a]	14.5	July 1977
	(3)	Support for Development Plan[a]	568.0	Nov. 1977
	(4)	Water and Sewerage[a]	28.6	Jan. 1978
Y.P.D.R.	(1)	Electric Power Project	19.8	July 1977
	(2)	Housing Development	14.2	July 1977
Bangladesh	(1)	Mining Development	50.0	July 1977
	(2)	Railway	50.0	July 1977
Brazil		Electric Power Project	55.0	Dec. 1977
Cameroon		Song-Loulou Dam	29.9	Feb. 1977
Gabon		Railway	20.6	Apr. 1978
Gambia		Airport	6.6	Feb. 1977
Ghana		Electric Power	29.9	Feb. 1977
India		Electric Power	100.0	June 1977
Indonesia		Highway	50.0	May 1977
Mali	(1)	Selingue Dam	15.0	Feb. 1977
	(2)	Livestock Project	13.3	Feb. 1978
Senegal		Highway	35.6	Jan. 1978
Taiwan		Telecommunications	29.2	Mar. 1978

Note: [a] It is not clear from quoted sources whether the specifically funded projects are part of the general support for the development plan.

Sources: *Middle East Economic Digest* (26 August, 9 December, and 22 December 1977); *Middle East Economic Survey* (7 March, 14 March, 13 June, 20 June, 27 June, 11 July, 18 July, 1 August, 8 August, 10 October, 28 November 1977, 23 January, 3 April, 8 May, 1978); *Petromoney Report* (8 July 1977); private correspondence with SAFD officials.

Notes

1. *Middle East Economic Survey,* 1 May 1978, pp. 11-12.
2. OECD, *Development Co-Operation,* p. 107.

BIBLIOGRAPHY

Askari, Hossein, Yves Bizien, and Ekram Hossain. "A Study of the Effectiveness of Devaluations and Revaluations in the Post-War Era." *Economia Internazionale.* Vol. XXVI (August-November, 1973).

Askari, Hossein and John Thomas Cummings. *Middle East Economies in the 1970s: A Comparative Approach.* New York: Praeger Publishers, 1976.

Askari, Hossein and Michael Glover. *Military Expenditures and the Level of Economic Development.* Austin: Bureau of Business Research, The University of Texas at Austin, 1977.

Brook, Ezriel, Enzo R. Grilli, and Jean Waelbroek. "Commodity Price Stabilization and the Developing Countries: The Problem of Choice." World Bank Staff Working Paper No. 262, International Bank for Reconstruction and Development (July 1977).

British Petroleum Company. *Statistical Review of the World Oil Industry.* London, 1976.

Diaz-Alejandro, Carlos F. *Less Developed Countries and the Post-1971 International Financial System.* Essays in International Finance, No. 108. Princeton, New Jersey, 1975.

Fallah, Reza and Fereidun Fesharaki. "The New World Order: The Energy Dimension." *OPEC Review* (October, 1977).

Food and Agricultural Organization. *Monthly Bulletin.* Rome, July/August 1976 and February 1977.

Houthakker, H. S. and S. P. Magee. "Income and Price Elasticities in World Trade." *Review of Economics and Statistics.* Vol. LI (May 1969).

International Currency Review (various issues).

International Monetary Fund. *Direction of Trade.* Annual Editions, 1969-1975 and 1970-1976, Washington, D.C.

__________. *Annual Report of the International Monetary Fund.* Washington, D. C., 1975, 1976, and 1977 issues.

__________. *International Financial Statistics.* Washington, D. C., March 1977.

__________. *IMF Survey.* Washington, D. C., 5 September 1977.

International Petroleum Encyclopedia. Tulsa, Oklahoma: Petroleum Publishing Co., 1976.

Johnson, Harry G. "Commodities: Less Developed Countries' Demands and Developed Countries' Response." Paper delivered to the Massachusetts Institute of Technology Workshop on Specific Proposals and Desirable DC Response to LDC Demands Regarding the New International Economic Order. Cambridge, Massachusetts, 17-20 May 1976.

Jeune Afrique. "L'Aide Arabe et le Financement du Developpement Africain." Lettre d'Information N.1 (January 1976).

Kindleberger, Charles. *International Economics*. 4th ed. Homewood, Illinois: Richard D. Irwin, 1968.

Kuwait Fund for Arab Economic Development. *Annual Report*, 1975 and 1976 issues.

Magee, Stephen P. "Tariff Preferences for Less Developed Countries, World Wage Income and Welfare." Background paper prepared for the United Nations—International Labor Organization World Conference on Employment, Income Distribution, and Social Progress in the International Division of Labor. New York, 1976.

Middle East Currency Reports (various issues).

Middle East Economic Digest (various issues).

Middle East Economic Survey (various issues).

Modigliani, Franco and Hossein Askari. *The Reform of the International Payments System*. Essays in International Finance No. 89. Princeton, New Jersey, 1971.

Organization for Economic Cooperation and Development. *Development Co-operation: 1976 Review*. Paris, November 1976.

Park, Y. S. *The Link Between Special Drawing Rights and Development Finance*. Essays in International Finance No. 100. Princeton, New Jersey, 1973.

Richardson, Harry W. *Economic Aspects of the Energy Crisis*. Lexington, Massachusetts: D. C. Heath and Company, 1977.

United Nations World Food Conference. *Assessment of the World Food Situation: Present and Future*. New York, 1974.

Willett, Thomas D. *The Oil-Transfer Problem and International Economic Stability*. Essays in International Finance No. 113. Princeton, New Jersey, 1975.

Williamson, John H. *The Crawling Peg*. Essays in International Finance No. 50. Princeton, New Jersey, 1965.

INDEX